List of

2	3	5							29
31	37	41							71
73	79	83							13
127	131	137			[illegible]	[illegible]	163	167	173
179	181	191	193	197	199	211	223	227	229
233	239	241	251	257	263	269	271	277	281
283	293	307	311	313	317	331	337	347	349
353	359	367	373	379	383	389	397	401	409
419	421	431	433	439	443	449	457	461	463
467	479	487	491	499	503	509	521	523	541
547	557	563	569	571	577	587	593	599	601
607	613	617	619	631	641	643	647	653	659
661	673	677	683	691	701	709	719	727	733
739	743	751	757	761	769	773	787	797	809
811	821	823	827	829	839	853	857	859	863
877	881	883	887	907	911	919	929	937	941
947	953	967	971	977	983	991	997	1009	1013
1019	1021	1031	1033	1039	1049	1051	1061	1063	1069
1087	1091	1093	1097	1103	1109	1117	1123	1129	1151
1153	1163	1171	1181	1187	1193	1201	1213	1217	1223
1229	1231	1237	1249	1259	1277	1279	1283	1289	1291
1297	1301	1303	1307	1319	1321	1327	1361	1367	1373
1381	1399	1409	1423	1427	1429	1433	1439	1447	1451
1453	1459	1471	1481	1483	1487	1489	1493	1499	1511
1523	1531	1543	1549	1553	1559	1567	1571	1579	1583
1597	1601	1607	1609	1613	1619	1621	1627	1637	1657
1663	1667	1669	1693	1697	1699	1709	1721	1723	1733
1741	1747	1753	1759	1777	1783	1787	1789	1801	1811
1823	1831	1847	1861	1867	1871	1873	1877	1879	1889
1901	1907	1913	1931	1933	1949	1951	1973	1979	1987
1993	1997	1999	2003	2011	2017	2027	2029	2039	2053
2063	2069	2081	2083	2087	2089	2099	2111	2113	2129
2131	2137	2141	2143	2153	2161	2179	2203	2207	2213
2221	2237	2239	2243	2251	2267	2269	2273	2281	2287
2293	2297	2309	2311	2333	2339	2341	2347	2351	2357
2371	2377	2381	2383	2389	2393	2399	2411	2417	2423
2437	2441	2447	2459	2467	2473	2477	2503	2521	2531
2539	2543	2549	2551	2557	2579	2591	2593	2609	2617
2621	2633	2647	2657	2659	2663	2671	2677	2683	2687
2689	2693	2699	2707	2711	2713	2719	2729	2731	2741
2749	2753	2767	2777	2789	2791	2797	2801	2803	2819
2833	2837	2843	2851	2857	2861	2879	2887	2897	2903
2909	2917	2927	2939	2953	2957	2963	2969	2971	2999
3001	3011	3019	3023	3037	3041	3049	3061	3067	3079
3083	3089	3109	3119	3121	3137	3163	3167	3169	3181
3187	3191	3203	3209	3217	3221	3229	3251	3253	3257
3259	3271	3299	3301	3307	3313	3319	3323	3329	3331
3343	3347	3359	3361	3371	3373	3389	3391	3407	3413
3433	3449	3457	3461	3463	3467	3469	3491	3499	3511
3517	3527	3529	3533	3539	3541	3547	3557	3559	3571
3581	3583	3593	3607	3613	3617	3623	3631	3637	3643
3659	3671	3673	3677	3691	3697	3701	3709	3719	3727
3733	3739	3761	3767	3769	3779	3793	3797	3803	3821
3823	3833	3847	3851	3853	3863	3877	3881	3889	3907
3911	3917	3919	3923	3929	3931	3943	3947	3967	3989
4001	4003	4007	4013	4019	4021	4027	4049	4051	4057

James M. McCanney, M.S.

- CALCULATE PRIMES -

DIRECT PROPAGATION OF THE PRIME NUMBERS

THE OLDEST PROBLEM IN MATHEMATICS HAS FINALLY BEEN SOLVED

BOOK and DVD SIMPLIFIED FOR THE GENERAL PUBLIC

___THE ORIGINAL MATHEMATICAL TREATISE IS INCLUDED AS AN APPENDIX___

Dedicated to my father, who could not afford to go to college, but who understood the value of education (even though he rarely knew what I was studying) and who taught me at an early age to seek the truth ... no matter what path I followed ... no matter what the cost.

Published By
jmccanneyscience.com press
Minneapolis, Minnesota

The official web site for this book is www.calculateprimes.com

*** Patent Pending – see Trade Mark Notices - TM**

1st Printing - 2007

ISBN 978-0-9722186-6-5

About the author and this text:

This book, as promised, will present to the public a rarely seen side of the author. James McCanney is known worldwide for his research and books as well as radio and lecture appearances dealing with the electrical nature of the cosmos. What many people do not know is that he is also a Mathematician who has worked on and solved some of the more complex problems of Physics and Mathematics, and is an expert in computer design, solid-state physics, telecommunications and computer protocols.

He has worked about half of his lengthy career in private industry, owning several of his own companies and also taught at the University level in Physics, Mathematics, Computer Science and Astronomy. Much of this was accomplished in multi-lingual settings, having worked in the USA, Latin America and with high-level Russian scientists. He has presented his findings at numerous international conferences and is a regular presenter at American Geophysical Union meetings. He has presented his theoretical research in locations such as Los Alamos National Laboratories, International Air shows and International Electric Propulsion conferences. Understanding his background is important in placing the current text into perspective.

James M. McCanney, M.S. received sound classical physics training at St. Mary's University (Winona, Minnesota) receiving a Bachelor of Arts degree with a double major in Physics and Mathematics in 1970. He was offered full scholarship awards to three major US physics graduate schools to pursue his graduate physics studies. However, he chose instead to postpone graduate studies for a period of three years while he traveled and taught Physics and Mathematics in Latin America.

During this time he spent a good deal of time traveling to ruins of ancient cities and archeological sites, studying firsthand many times as the ruins were dug from under dirt that had not been moved for thousands of years. Also during this time he developed the basis for his theoretical work that would, at a later date, deal with the celestial mechanics of N-bodies and plasma physics, as well as the groundwork for the current treatise on prime numbers.

He has stated that in his travels he learned that nature provides elegant yet simple solutions to even the most complex problems, and that she reveals these fundamental secrets to only those who pursue truth for truth's sake, without the encumbrances of modern life or economic rewards that so many times cloud the visions of humans, especially scientists, most of whom are entrenched in the age of big business and government monopolized science.

With this new understanding of archeology, astronomy of the ancients, physics and the world around him, Mr. McCanney returned to graduate school in 1973 and earned a Master of Science degree in nuclear and solid-state physics from Tulane University, New Orleans, LA. His research involved a National Institute of Health research grant (in conjunction with Gulf South Research Institute) to study visible light and x-ray sensitive photo polymers to be used in medical Xerox x-ray machines. The results of his research resulted in patented discoveries. He was again offered a full fellowship to continue on with Ph.D. studies, but once again he declined and returned to Latin America to study archeology and teach Physics, Mathematics and Computer Science in Spanish. He continued his personal work to explore the mysteries of celestial mechanics and its relationship to the planets, moons and other celestial bodies.

At this time a fellow Ph.D. at Tulane wrote in a letter to a professor at another University, "the first observation that comes to mind concerning Jim McCanney is his unusually broad range of scientific and intellectual interests. His mind is a veritable fund of information and ideas on topics as varied as Egyptian pyramids and the classification of mushrooms. His character and morals are of the highest caliber. He possesses an excellent aptitude for design and construction of research and measurement systems as well as a creative bent in understanding and explaining observed physical processes. He is able to engage in significant research with little or no direct supervision. His research experience is such that he could involve undergraduates with minimal outlay of funds. He has the ability to command respect of those in his classes and is able to generate an unusually good rapport with students."

After four years living and teaching at the university level in Latin America, in 1979 Mr. McCanney joined the faculty of Cornell University, Ithaca N.Y. as an introductory instructor in physics. It was during this time that he had access to NASA data returning daily from the Voyager I and II spacecraft as they traveled by the planets Jupiter, Saturn and beyond (as well as data from many other space craft). It was here he recognized that the mathematical models in his theoretical work regarding the electro-dynamic nature of the solar system and universe had its signatures in the new data that was streaming in from the edges of the solar system.

His papers were published at first in the peer reviewed astrophysical journals, but soon he began to receive resistance from the standard astronomical community. Mr. McCanney was removed from his teaching position because the standard scientific community did not accept his beliefs regarding the electro-dynamic nature of the solar system, beliefs that are just now becoming the basis for modern Space Science and Astrophysics.

His innovative theories on plasma physics and a new model for fusion in the solar atmosphere provided the basis for the electric fields and plasma discharge phenomena that have become the core elements of his theoretical models of the true nature of the solar system in which we live, and which are the fast becoming the basis for every international conference on space weather and space physics.

Upon being removed from the physics department for his then radical beliefs, Mr. McCanney was rehired shortly thereafter by the mathematics department also at Cornell University, where he taught for another year and a half and continued to publish his papers in peer reviewed astrophysical journals. Again astronomers coerced his removal from the Mathematics Department and ultimately he was blackballed from publishing in the astrophysics journals in1981.

His first peer-reviewed paper in 1980 "Continuing Galactic Evolution" (written while he was in the Physics Department but published while in the Mathematics Department of Cornell University) advanced a new technique for solving simultaneous differential equations for the orbits of stars in a galaxy. It is said that only a small handful of people in the world understood the significance of this short but important paper. With a second paper "Saturn's Sweeper Moons Predicted", he was, according the department chairman at the time, the only Cornell Mathematics Department faculty member ever to be published in the peer reviewed astrophysics journals.

During this time Mr. McCanney established himself as the originator of the theoretical work regarding the electrical nature of the cosmos for which he coined the term "THE ELECTRIC UNIVERSE", which today is being proven correct on an ongoing basis by space-probes returning data from outer space. Many of his predictions such as x-rays to the sunward side of comet nuclei, that comet nuclei would be found to have no ice or water frozen on their surfaces and that comets interact electrically with the Sun as well as solar electrical activity affecting Earth weather, have now been confirmed by direct measurements in 1986, 1996, 2001 and 2002 to the present day. Many other more abstract concepts have also been verified.

After the Cornell years Mr. McCanney worked in the telecommunications industry as a Principle Systems Engineer at a major telecommunications corporation designing computer telecommunications equipment, and also ran a number of his own companies dealing with satellite communications and electrical propulsion systems. He attended and spoke at numerous international conferences on the future of space flight using his designs of electro-magnetic propulsion.

It was during these same years that he continued to interact with the scientists at Los Alamos Labs and Goddard Space Flight Center during the era of the first comet fly bys of comets Giacobini-Zinner and Comet Halley. Throughout the years and the decades, he has always remained an Independent Scientist, never using government funding or positions to further his personal research.

In the mid-1990s Mr. McCanney's work was recognized by a group of high-level Russian scientists at the University at Novosibirsk who had measured but did not understand electro-dynamic effects around Earth and in the solar system. Upon discovering Mr. McCanney's work, they translated all of his papers to date into Russian. These are still being taught at the university level as the leading edge of research in this field. It is only due to the ongoing and intentional efforts of NASA that his work has received such little attention in the western scientific community and press.

In the background Mr. McCanney continued working on the illusive mathematical problems facing celestial mechanics, that is, the unknown method of solving simultaneous differential equations. He believed that nature had provided a simple but elegant method to solve these complex problems that did not exist in current mathematical capabilities. He envisioned a mathematical technique that would be applied and reapplied to generate a final solution. In this respect, many unsolved problems of mathematics and quantum mechanics and even genetic molecular structures would have their solutions in this new mathematical technique.

His search lead to a technique that is the basis for the discovery of the solution to the prime number problem presented in this book, and which now holds promise at solving some of the great mysteries of Quantum Physics as well as Genetics. It is based on the "Mathematical Operators" that the Professor calls "Generator Functions".

These mathematical expressions operate on fundamental units or building blocks of nature such as the numbers 0 and 1 (as in the case of the prime number solution in this book) or atomic nuclei (in the case of quantum mechanics) or atoms and small molecules (in the construction of large organic molecules or genetic structures). According to Mr. McCanney, "the Generator Function is an operator that reflects conditions in nature, it builds up a first layer when applied to the basic building blocks, and then is applied to this construct to form a second more complex structure. The same Generator Function is then reapplied to this result and the process continues, with each successive application modifying the prior set of results".

This leads to questions that are being asked for the first time. For example in genetics, could the makeup of the human biped be not only duplicated somewhere else in the universe, but is the norm rather than the exception based on "Generator Functions" that operate on the building blocks of life. Could this pattern be replicated on the far side of the universe just as a diamond would appear the same in such a distant place?

On a simpler note, as the prime numbers would be discovered by intelligent beings on the far side of the universe, they would follow the same mathematical Generator Function that we would discover here on earth. The implications of this new mathematical technique may someday very well overshadow this initial application and the discovery of the formula for directly calculating the prime numbers to infinity as presented in this book and DVD lecture (Appendix I contains the original mathematical treatise).

An essential aspect of the Generator Function concept is that it creates intermediate or "false" solutions that are necessary steps to generate the final correct solution. This is possibly but one of the reasons why so many mathematicians have worked around the prime number problem without finding a solution, as they failed to see the total picture as presented in this book. Countless mathematicians have come to the conclusion that that prime numbers are random with but passing patterns. This book proves that to be incorrect. When you finally look at the prime number table and see why every number is there and how it is related to all the other primes, with all of their "ancestors" bridging back to the numbers 0 and 1, you will understand the powerful impact this book will have on the future.

This is a brief but incomplete history of the background that makes "The Professor", as many people call Mr. McCanney, a person with a unique background and qualifications.

Mr. McCanney airs a weekly radio show "The James McCanney Science Hour – At the Crossroads" heard world wide on short wave radio and also broadcast on the internet. He has been a popular regular guest on many national and international radio shows. Most people know of "The Professor" from his work regarding the electrical nature of outer space and its relation to our planet and human beings that inhabit our blue planet.

He has worked over his career on the current topic of the prime number solution and has taught the following mathematics courses at the University level (in addition to Physics, Computer Science and Astronomy); Abstract Algebra, Linear Algebra, Matrix Algebra, Probability and Statistics, Statistics for Computing, Mathematical Logic, Theory of Numbers, Calculus I, II and III, Engineering Math I and II, Advanced Topics in Geometry and Topology.

He has been an avid student of the history of both Physics and Mathematics and has studied the past to understand what is of greatest importance in the future of these fields of study. The present book is a glimpse into this rarely seen side of Professor McCanney. He has stated that his greatest discoveries to date include the cause of surface fusion on the sun, which gives rise to the solar electric field that drives the electrical nature of the solar system. His second great discovery, which was published almost 2 decades before its discovery, was the prediction that comets would produce x-rays far to the sunward side, as well as the electrical nature of comets and the prediction that comet nuclei would not contain water or ice, but would be hot dry asteroid type bodies (a prediction that has repeatedly been proven by subsequent space probes), much to the surprise of professional astronomers.

A third major theoretical discovery, which has been observed both in nature and subsequently measured in the laboratory, he calls "The Induced Electric Dipole Red Shift". This is an effect on photons of light moving through a non-uniform electric field that causes both a red shift as well as a bending of light. It also explains the previously unknown cause of pair production (electron-positron production) in photons passing near heavy atomic nuclei. His theoretical work then restructures the laws of cosmology based on this new form of red shift.

And last but not least, his work has rediscovered the work of Tesla and further provides a theoretical basis for deriving unlimited electrical energy from the ionosphere. Another result of his decades of research has been the realization that severe earth weather including hurricanes and tornados are powered by electrical discharges from the ionosphere to the clouds in storm systems. The list of scientific discoveries also includes an electromagnetic propulsion system for use in space travel that has been field-tested.

In spite of these discoveries that were confirmed in many cases decades after Mr. McCanney's theoretical predictions were published, he now has stated that the current treatise on prime numbers potentially may have the greatest significance in the history of both Physics and Mathematics. The solution of the Prime Number Problem has eluded every great mind since the Greeks first defined it nearly 2500 years ago. Cauchy, Gauss, Riemann, Euler, and countless others spent their entire lives scouring this problem, yet could not crack the fundamental problem of being able to directly calculate the prime numbers. Countless major unsolved problems in both Physics and Mathematics base their solutions on the prime numbers, yet no one could directly calculate these illusive and perplexing numbers, that is, until now.

It was in quest of this solution and the understanding of its importance that has brought us to herald this work, which is now explained in a way that only the good Professor can do, making it understandable by anyone with a rudimentary knowledge of mathematics. The solution is simple and elegant, the way nature intended.

At present there are two more books destined to become part of this series that will be released at later dates.

a.s.

TABLE OF CONTENTS

PREFACE

Putting this book into perspective

Preface by the Author

The solution to the prime number problem haunted me from my earliest introduction to it in late grade school or early high school. I had worked on trying to find patterns and certainly found myself "discovering" many of the same facets of the prime numbers and following many of the same false leads that the hundreds of mathematicians had stumbled upon over the past 2500 years.

In pursuit of a solution over many decades, I read everything that existed and got to know the paths of great mathematicians such as Gauss, Euler and Reimann, the same great names that brought us so many unbelievably complex advancements in every field of mathematics. Certainly the solution to understanding the primes had its base in the foundations put down by these greatest of mathematicians. My efforts of course came and went as I worked on other pursuits. I had to leave my prime number file on the shelf to collect dust only to return time and time again to take up the battle.

Then finally a number of years ago I sat and stared into a mental mirror and had to come to task with a fundamental issue. If all of these great mathematicians could not solve this problem, were they all wandering perhaps down the wrong roads? What were they all missing? I decided at that point on two issues, both of which later proved to be correct. The first was something I had discovered while wandering in the rain forests of Central America and camping in the remotest wilderness areas of northern Canada. This was the realization that the solution to understanding the prime numbers was like everything else in nature ... it most certainly did exist ... and it had to be elegant but simple. And secondly, there had to be a new "method" of mathematics that had not yet been discovered.

As I later found out, I was correct on both counts. So it was at that fateful point in my theoretical life that I made a conceptual decision that was the equivalent of cutting off my right arm. I would take indefinite leave of working on the project labeled "Prime Number

Problem" ... until such time as I had forgotten all of what had to be false leads of all these other mathematicians I had studied and likewise abandoned my own set of false leads.

Many times I had to resist the temptation to reopen the "Prime Number Problem" file as I saw it sitting on the shelf, since I had not shed myself of all the dead weight yet. Patience my son, patience a little voice said inside of my head. And I then continued with the many other pursuits that have kept me more than busy over the years.

Then one day, after possibly 2 years of not thinking about the prime number problem, I sat down with pen and paper and the solution starting flowing like water over a newly opened reservoir. I was setting on a deck in the warm sun, and with a fresh supply of paper, began putting together the pieces of the puzzle. In retrospect, at times I thought that there was one or another aspect of the total solution that was paramount or more important than any other, but in the final analysis there are so many subtle and intricate solutions that have to be woven together to make the final solution work, that it is not possible to weight any one aspect above the others. As with many events, it was a collection of factors all working in harmony to give the true vision of the prime numbers.

Little did I know that my brains had been purged of "the past prime number problem" information and the fresh look was truly what was needed. As I sat down that morning with pen in hand, I began probing into the middle of the prime number table. It hit me like a ton of bricks and I did not stop working until the entire solution method had been worked out ... it was so logical and so simple ... and so elegant ... I had solved the initial portion ... the essence of the prime number problem. It took another 6 weeks of daily work to iron out the details and put it into a form that other people would understand. Additionally I had to answer for myself all of the difficult questions I knew that professional mathematicians would ask when they saw the formula that directly calculates the prime numbers. Most of all, I had discovered a new mathematical technique that I knew had to exist and may possibly be used on other more complex mathematical problems.

This book has been designed for the general public and also for mathematicians who want an overview of the solution to the oldest and most illusive mathematical problem in the history of modern man.

jim mccanney

I. THE ANCIENT GREEKS AND THE PRIMES

This book will not attempt to be a resource for the history and drama regarding the prime numbers. There are many good sources on the Internet (as always beware of the few quack sites ... the good sites are based usually at universities and cross reference the other good web sites). Also there are many good books written by professional mathematicians regarding this topic (Appendix II of this book gives a few references based on level of difficulty).

The ancient Greeks are credited with the advancement of knowledge of the complexity of prime numbers. Prime numbers of course are those numbers that only have themselves and the number 1 as factors. To test whether a number is prime or not, based on this definition, requires one to laboriously divide the number by every number smaller than the given number. If all of the divisions or "factorizations" end up with a remainder, then there are no factors and the number is added to the list of known prime numbers. There are of course many tricks to reduce the number of computations. But to date, the basic process remains the same.

The Greek number system, as with most old systems of counting, depended on placing objects (apples or soldiers ... what ever they were counting) into rows and columns. Thus the number 20 for example could be represented in any of the following patterns.

```
XX   OR  XXXX  OR  XX   XX
XX       XXXX      XX   XX
XX       XXXX      XX   XX
XX       XXXX      XX   XX
XX       XXXX      XX   XX
XX
XX
XX
XX
XX
```

But, take away just one element to make 19, and there is no way to arrange the apples or soldiers in neat sets of rows and columns. This seems fairly rudimentary; yet this dilemma has plagued mathematicians since the time of the ancient Greeks. That is, how can you predict which numbers belong to this unusual group known as "The Primes Numbers"? What this also illustrates is that the prime numbers can be represented by geometrical patterns, the same patterns as are found many times in nature.

The Greeks had other classifications for numbers such as "perfect numbers", which referred to numbers that have their many factors also adding up to equal the number. For example 6=1x2x3 and also 6=1+2+3, so 6 is a "perfect number". They recognized many of the properties that defined the "Theory of Numbers" problems that have kept mathematicians busy for the past 2500 years.

They were impressed with infinity for example, yet it was not until a modern mathematician named George Cantor worked out the details and solidified the topic. Thanks to Cantor we now have a firm grasp of the concept of infinity. The Greeks had many other ideas that held fast for thousands of years. For example Aristotle taught that objects would come to a stop by their nature, not understanding the principle of friction. Isaac Newton reversed that concept with his three laws of motion. Likewise Galileo reversed the Aristotelian concept that larger heavier balls fell faster with his famous experiment at the leaning tower of Pisa, although the standard scientists of the time refused to believe this and continued teaching Aristotle's concepts for a long time after Galileo brought in the age of experimentation.

The ancient Greeks knew that the world was round and actually made primitive measurements. Yet this knowledge seemed to have been lost at least amongst the common people until well into the middle ages. Yet other ideas such as comets being fire in the atmosphere and the earth being the center of the universe held fast until Kepler and Copernicus discovered otherwise.

Regarding the prime numbers, however, every great mathematician since the time of the ancient Greeks and throughout history have wrestled throughout their lifetimes with this seemingly simple problem, failing to discover a formula that could directly predict the prime numbers. Over 100 years ago mathematicians gave up the effort and resorted to higher mathematics to determine the "density of primes", that is, they developed complex equations to determine in general how many prime numbers there were in a certain region of the number line. The most famous of these is known as "The Riemann

Hypothesis" and is touted today as the Greatest Unsolved Problem in Mathematics. Hundreds of proofs in both Physics and Mathematics depend on its correctness, as these begin by assuming that it must be true and then proceed with further developments. Even with modern super computers finding very large prime numbers, mathematicians cannot be sure until there is a generalized mathematical proof.

A number of years ago one result claimed to have proven that the prime numbers are random and that there would be no formula for their prediction. Mathematicians were relieved in a sense. This book, however, shows that not only is there an equation to directly calculate the prime numbers using only addition and subtraction, but it also shows that there are amazing patterns, mathematical properties and associations amongst the prime numbers. Furthermore, these patterns behave like waves in a pond. It also shows that there are properties such as symmetry, where the prime numbers are symmetrically located around certain "magic numbers", and that groups of prime numbers can generate entire other groups of primes, that in turn can generate the original group of primes when the same equation is applied to the second group of primes. This property is called "reciprocity".

When you see and understand the solution described in this book, and understand what I call "the curtain" that has blocked understanding of this throughout history, you will see that all of these mathematicians worked around this solution repeatedly ... but never saw through the thin layer that separated them from solving this perplexing problem. There are many subtleties. For example, one has to generate certain "false" solutions, which coexist with the good solutions (the "true" primes), before the final solution is achieved.

With every application of what I call "The Generator Function", prime numbers are not generated one at a time, but in groups. This group is then used to generate a subsequent group that will identify even more prime numbers. Each time the Generator Function is applied it generates an infinite number of future "prime candidates" in waves that act as waves in a pond, repeating with a set "wave length" all the way to infinity. This wave starts at a certain point on the number line ... at one of the so called "magic numbers" ... and not only proceeds to infinity, but also moves to the left, or in the negative direction. These subtleties will all become clear when actually using this process to directly calculate the prime numbers.

II. ABOUT THE INCLUDED DVD LECTURE

There is a DVD lecture included in the rear pouch of this book. The DVD is copyrighted and the material patented. The DVD cannot be copied, reproduced or resold. It is a part of this book.

When I first started releasing the information in the original mathematical treatise (contained in Appendix I of this book) and while doing test runs to make sure it was clearly written and understandable by well trained mathematicians and scientists, it soon became apparent that there are so many subtleties and new aspects to this work, that even well trained mathematicians needed coaching to put all the pieces together. The DVD lectures also allow me to divert from the basic text and add material visually to aid people not well versed in mathematics. I firmly believe that the concepts here are so basic that nearly everyone can understand them, given a good visual presentation.

So attempting to present this material to the average person on the street, as this book is doing, would be impossible without the special aid of the DVD and visual lecture. It also is a more palatable format for many to watch to get the ideas without the labor of reading this text, although you should first review the DVD and then read the book that does contain additional material not covered in the DVD. The DVD also facilitates teachers who want to present this material to classrooms of students and was designed for everyone from 4th grade to graduate schools and professional mathematicians and scientists with the 2 part presentation … "Basic" and "Advanced" topics.

Note that there are long lists of advanced topics not contained in this book and DVD that will appear in other venues and formats. The current topic only deals with the basics of directly calculating the prime numbers from the basic starting point of the numbers 0 and 1.

The next chapter lists the prime numbers to 9109. Remove and photocopy these pages; enlarge up to at least 8 ½" x 11"; make several copies as you will need these to make notes and do exercises as we discuss the prime generation process, demonstrate patterns, etc..

ALSO PLEASE … take one very long and hard last good look at this seemingly random list of numbers, since they will never look the same to you again.

III. THE PRIME NUMBER TABLE

List of Prime Numbers to 9109

(Photocopy about 5 copies onto full size paper to perform exercises)

2	3	5	7	11	13	17	19	23	29
31	37	41	43	47	53	59	61	67	71
73	79	83	89	97	101	103	107	109	113
127	131	137	139	149	151	157	163	167	173
179	181	191	193	197	199	211	223	227	229
233	239	241	251	257	263	269	271	277	281
283	293	307	311	313	317	331	337	347	349
353	359	367	373	379	383	389	397	401	409
419	421	431	433	439	443	449	457	461	463
467	479	487	491	499	503	509	521	523	541
547	557	563	569	571	577	587	593	599	601
607	613	617	619	631	641	643	647	653	659
661	673	677	683	691	701	709	719	727	733
739	743	751	757	761	769	773	787	797	809
811	821	823	827	829	839	853	857	859	863
877	881	883	887	907	911	919	929	937	941
947	953	967	971	977	983	991	997	1009	1013
1019	1021	1031	1033	1039	1049	1051	1061	1063	1069
1087	1091	1093	1097	1103	1109	1117	1123	1129	1151
1153	1163	1171	1181	1187	1193	1201	1213	1217	1223
1229	1231	1237	1249	1259	1277	1279	1283	1289	1291
1297	1301	1303	1307	1319	1321	1327	1361	1367	1373
1381	1399	1409	1423	1427	1429	1433	1439	1447	1451
1453	1459	1471	1481	1483	1487	1489	1493	1499	1511
1523	1531	1543	1549	1553	1559	1567	1571	1579	1583
1597	1601	1607	1609	1613	1619	1621	1627	1637	1657
1663	1667	1669	1693	1697	1699	1709	1721	1723	1733
1741	1747	1753	1759	1777	1783	1787	1789	1801	1811
1823	1831	1847	1861	1867	1871	1873	1877	1879	1889
1901	1907	1913	1931	1933	1949	1951	1973	1979	1987
1993	1997	1999	2003	2011	2017	2027	2029	2039	2053
2063	2069	2081	2083	2087	2089	2099	2111	2113	2129
2131	2137	2141	2143	2153	2161	2179	2203	2207	2213
2221	2237	2239	2243	2251	2267	2269	2273	2281	2287
2293	2297	2309	2311	2333	2339	2341	2347	2351	2357
2371	2377	2381	2383	2389	2393	2399	2411	2417	2423
2437	2441	2447	2459	2467	2473	2477	2503	2521	2531
2539	2543	2549	2551	2557	2579	2591	2593	2609	2617
2621	2633	2647	2657	2659	2663	2671	2677	2683	2687
2689	2693	2699	2707	2711	2713	2719	2729	2731	2741
2749	2753	2767	2777	2789	2791	2797	2801	2803	2819
2833	2837	2843	2851	2857	2861	2879	2887	2897	2903
2909	2917	2927	2939	2953	2957	2963	2969	2971	2999
3001	3011	3019	3023	3037	3041	3049	3061	3067	3079
3083	3089	3109	3119	3121	3137	3163	3167	3169	3181
3187	3191	3203	3209	3217	3221	3229	3251	3253	3257
3259	3271	3299	3301	3307	3313	3319	3323	3329	3331
3343	3347	3359	3361	3371	3373	3389	3391	3407	3413
3433	3449	3457	3461	3463	3467	3469	3491	3499	3511
3517	3527	3529	3533	3539	3541	3547	3557	3559	3571
3581	3583	3593	3607	3613	3617	3623	3631	3637	3643
3659	3671	3673	3677	3691	3697	3701	3709	3719	3727
3733	3739	3761	3767	3769	3779	3793	3797	3803	3821
3823	3833	3847	3851	3853	3863	3877	3881	3889	3907

3911	3917	3919	3923	3929	3931	3943	3947	3967	3989
4001	4003	4007	4013	4019	4021	4027	4049	4051	4057
4073	4079	4091	4093	4099	4111	4127	4129	4133	4139
4153	4157	4159	4177	4201	4211	4217	4219	4229	4231
4241	4243	4253	4259	4261	4271	4273	4283	4289	4297
4327	4337	4339	4349	4357	4363	4373	4391	4397	4409
4421	4423	4441	4447	4451	4457	4463	4481	4483	4493
4507	4513	4517	4519	4523	4547	4549	4561	4567	4583
4591	4597	4603	4621	4637	4639	4643	4649	4651	4657
4663	4673	4679	4691	4703	4721	4723	4729	4733	4751
4759	4783	4787	4789	4793	4799	4801	4813	4817	4831
4861	4871	4877	4889	4903	4909	4919	4931	4933	4937
4943	4951	4957	4967	4969	4973	4987	4993	4999	5003
5009	5011	5021	5023	5039	5051	5059	5077	5081	5087
5099	5101	5107	5113	5119	5147	5153	5167	5171	5179
5189	5197	5209	5227	5231	5233	5237	5261	5273	5279
5281	5297	5303	5309	5323	5333	5347	5351	5381	5387
5393	5399	5407	5413	5417	5419	5431	5437	5441	5443
5449	5471	5477	5479	5483	5501	5503	5507	5519	5521
5527	5531	5557	5563	5569	5573	5581	5591	5623	5639
5641	5647	5651	5653	5657	5659	5669	5683	5689	5693
5701	5711	5717	5737	5741	5743	5749	5779	5783	5791
5801	5807	5813	5821	5827	5839	5843	5849	5851	5857
5861	5867	5869	5879	5881	5897	5903	5923	5927	5939
5953	5981	5987	6007	6011	6029	6037	6043	6047	6053
6067	6073	6079	6089	6091	6101	6113	6121	6131	6133
6143	6151	6163	6173	6197	6199	6203	6211	6217	6221
6229	6247	6257	6263	6269	6271	6277	6287	6299	6301
6311	6317	6323	6329	6337	6343	6353	6359	6361	6367
6373	6379	6389	6397	6421	6427	6449	6451	6469	6473
6481	6491	6521	6529	6547	6551	6553	6563	6569	6571
6577	6581	6599	6607	6619	6637	6653	6659	6661	6673
6679	6689	6691	6701	6703	6709	6719	6733	6737	6761
6763	6779	6781	6791	6793	6803	6823	6827	6829	6833
6841	6857	6863	6869	6871	6883	6899	6907	6911	6917
6947	6949	6959	6961	6967	6971	6977	6983	6991	6997
7001	7013	7019	7027	7039	7043	7057	7069	7079	7103
7109	7121	7127	7129	7151	7159	7177	7187	7193	7207
7211	7213	7219	7229	7237	7243	7247	7253	7283	7297
7307	7309	7321	7331	7333	7349	7351	7369	7393	7411
7417	7433	7451	7457	7459	7477	7481	7487	7489	7499
7507	7517	7523	7529	7537	7541	7547	7549	7559	7561
7573	7577	7583	7589	7591	7603	7607	7621	7639	7643
7649	7669	7673	7681	7687	7691	7699	7703	7717	7723
7727	7741	7753	7757	7759	7789	7793	7817	7823	7829
7841	7853	7867	7873	7877	7879	7883	7901	7907	7919
7927	7933	7937	7949	7951	7963	7993	8009	8011	8017
8039	8053	8059	8069	8081	8087	8089	8093	8101	8111
8117	8123	8147	8161	8167	8171	8179	8191	8209	8219
8221	8231	8233	8237	8243	8263	8269	8273	8287	8291
8293	8297	8311	8317	8329	8353	8363	8369	8377	8387
8389	8419	8423	8429	8431	8443	8447	8461	8467	8501
8513	8521	8527	8537	8539	8543	8563	8573	8581	8597
8599	8609	8623	8627	8629	8641	8647	8663	8669	8677
8681	8689	8693	8699	8707	8713	8719	8731	8737	8741
8747	8753	8761	8779	8783	8803	8807	8819	8821	8831
8837	8839	8849	8861	8863	8867	8887	8893	8923	8929
8933	8941	8951	8963	8969	8971	8999	9001	9007	9011
9013	9029	9041	9043	9049	9059	9067	9091	9103	9109

IV. COPYRIGHTS - PATENTS AND TRADE MARKS

Due to the nature of the information presented in this book and DVD, a good part of the preparation for release of this book was in the form of legal work. Copyrights protect the written page and have a certain application in the protection and control of a work. They do not however protect conceptual ideas, inventions or terms. This is where Patents and Trademark Registrations are invoked.

The material in this book holds a Copyright, as does the original mathematical treatise in Appendix I. In addition to the Copyrights, Patent Pending status applies to all materials presented here and various terms have been registered with Trade Mark ™ status. There will undoubtedly be other registrations that proceed as this book is released and afterwards. **These issues will not be listed in this book** as such issues migrate and change with time. What is current today regarding these topics will possibly have additions and therefore this book is not an appropriate venue to list such information.

The professional staffs outside the scope of this book will handle issues regarding Copyrights, Patents or Trade Marks. The official web site for this book is **www.calculateprimes.com**

TEACHERS AND EDUCATORS – for K-12 only

In an effort to facilitate K-12 education and the propagation of this information to students, this material may be reproduced for the sake of your personal classroom educational purposes only, and for students ... especially those students who would otherwise not be able to afford these materials. These reproduced materials cannot be resold or redistributed. Any reproductions for education can only be used within the immediate physical classroom of a teacher. The DVD can NOT be duplicated.

PROHIBIT POSTING ON THE INTERNET

No part of the material in this book and/or DVD may be posted on the Internet in any form including transcriptions, paraphrasing or other forms without the express written permission of the legal staff at jmccanneyscience.com press.

V. DISCOVERY OF "THE MAGIC NUMBERS"

For the sake of simplicity for the common reader, some of the mathematical terms used in Appendix I (the Original Mathematical Treatise) will be changed. Part of the discovery of the process involved in directly calculating the prime numbers was the identification of certain important numbers. These are numbers that are pivotal to the symmetrical generation of the prime numbers. In Appendix I they are called "Sequential Prime Products", but we will just call them "Magic Numbers".

We will start with some basic examples to get you thinking in the mode of directly calculating prime numbers using addition and subtraction, but be aware that this will get somewhat more complicated later in the text, and I caution mathematicians to wait until the true definitions are listed, since the following examples will have flaws that will be corrected later in the text. The following examples are simply employed to get you thinking in the mode of directly calculating the prime numbers using just addition and subtraction.

To get started, let's jump into the middle of your prime number table and start with an exercise (please have your photocopied pages of pages 5 and 6 in front of you, and have pens with 2 or 3 different colored inks if possible). Find the numbers 29 and 31 on your sheet of prime numbers. We will start with a number between these two prime numbers that will have some very special properties. It is the number 30. We will see later why 30 is a "magic number". Write the number 30 on your sheet in red ink between the numbers 29 and 30.

Start by adding and subtracting 1 to 30. You will get two prime numbers 29 and 31 (check your list). Mark a small red "-1" by 29 and a small "+1" by the 31 in your table to signify that they were generated by adding and subtracting 1 to 30. Red is the color of the group of primes generated from the "magic number" 30. Now add and subtract 7 from 30, you get 23 and 37 (mark these with "-7" by 23 and "+7" by 37. The mathematical symbol for "add and subtract" is $\pm$. Now try 30 $\pm$ 11; this gives 19 and 41 (mark them with "-11" and "+11" respectively. Now try 30 $\pm$ 13. This gives 17 and 43 (mark them). Now try 30 $\pm$ 17. This gives 13 and 47 (mark them). Apart from our magic number 30, all the numbers we have worked with above were prime numbers found on your list of primes. We added and subtracted prime numbers to the magic number 30 to generate other primes.

If we proceed to 19, the next prime number, and attempt to add it to 30, we run into a little problem. 30 – 19 gives 11 which is a prime number, however 30 + 19 gives 49 which is NOT a prime number. The number 49 has factors of 7 x 7 and is not prime. This would stop most mathematicians from continuing. The process seems to break down. BUT, there will be something very important about the number 49 (and many other numbers like it). Even though it is not a prime itself, in the Generator Function that will be defined in a later chapter, 49 will be needed to generate certain "true" prime numbers and then it will be eliminated in a natural process ... never to resurface again.

Here again is another subtle but essential element of the process of directly calculating the prime numbers. One has to generate many "false primes" to get to the real primes. And in fact they are only a few steps away from the true list of real or "true" prime numbers. So close but yet so far ... that not a single mathematician throughout history saw through this thin curtain that hid this elegant but ever so simple process.

Observe two basic results. First, all of the resulting numbers generated above are prime numbers. We generated them by adding and subtracting other prime numbers to our "magic number" 30. Secondly, these numbers were generated in pairs, but not just any pairs ... they are symmetrical around the "magic number" 30. And last but not least they were all generated in a group related to the magic number 30.

By this point many professional mathematicians are already balking and imagining that this process will break down as we get to larger and larger numbers ... and it will in fact do just that. But we have laid down a simple foundation that needs a little help and we will then be able to describe "The Generator Function" that in fact will be used in a similar process to directly calculate the entire list of prime numbers in groups all the way to infinity.

Part of my realization that there were infinitely many "Magic Numbers" and the rhythms they possessed was just one of the aspects of the total solution of the prime number problem. As noted before, every great mathematician throughout history looked at some of these same results in part or from another point of view, not completely putting together all of the necessary pieces of the puzzle to come to the final and complete solution.

We will now continue with another example for the sake of illustration. Find the numbers 887 and 907 in your table of primes. Between these numbers is the number 900, which is not in your table (it is not a prime number), but is a multiple of the magic number 30. Write the number 900 between 887 and 907 in your table and proceed to look at the prime numbers to the right and left of 900 and mark the differences between these numbers and 900. To the left we have 900 – 887 = 13, 900 – 883 = 17, 900 – 881 = 19, etc. To the right of 900 you add 900 + 7 = 907, 900 + 11 = 911, 900 + 19 = 919, etc.

Sooner or later there will be some problems with this process, but we will show how to deal with these in the final form of the "Generator Function". Let us take one more example before starting this process from just 0 and 1 and generating the prime numbers using only addition and subtraction.

Find the numbers 2309 and 2311 in your prime number table. Write the number 2310 between them and proceed once again to the left and right of 2310 to take the difference between the prime numbers and 2310. For example 2310 - 1 = 2309, 2310 – 13 = 2297, 2310 – 17 = 2293, etc. The list of differences will be as follows;

–1, -13, -17, -23, -29, -37, -41, -43, -59, -67, -71, -73, -79, -87 etc

Now proceed in the positive direction from 2310. 2310 + 1 = 2311 2310 + 23 = 2333, 2310 + 29 = 2339, etc. The list of numbers added to 2310 to generate the prime numbers in the prime number table are

+1, +23, +29, +31, +37, +41, +47, +61, +71, +73, +83, etc.

Note that all of these numbers are prime numbers that are added to 2310 to generate more prime numbers, but eventually this would break down if you continue. We will see some of these examples later in the book. Note also that we subtracted 13 to get the prime number 2297, but *adding* 13 to 2310 did not generate a prime number ... 2310 + 13 = 2323 is a "false" prime. There is a subtle nature to the symmetry in the primes, and once again this is where the "false" primes come into play. They in fact constitute a subtle but essential part of the symmetry that is needed to generate the final list of prime numbers, but as you will see, the false primes become eliminated from the list when using the "Generator Function" that is defined later in the text. All of these examples are preparing you for that definition. Without this preparation most people including trained mathematicians would have a difficult time understanding the concepts and implications.

The number 2310 is another "magic number". As an exercise try working from multiples of 2310 (e.g. 2 x 2310 = 4620). What do you see? How do the patterns of prime numbers around 4620 compare to those around 2310 that we calculated above?

The following is a list of "Magic Numbers". These will actually be derived later in the book and you will not only understand where they come from, but why in fact they are magic numbers. There is an infinite number of magic numbers and they will be generated as the prime numbers are generated with the Generator Function. Every time a new set of prime numbers is generated by applying the Generator Function to an already existing list of "true" primes and associated "false" primes ... a new "magic number" will automatically be generated also.

Note that each magic number is a product of prime numbers. In the Original Mathematical Treatise in Appendix I these are called "Sequential Prime Products" as a new one is generated every time a new prime number is discovered. The next prime number in our growing list is multiplied by the previous "sequential prime product" or "magic number" to generate the subsequent one. The column in the middle of the table below represents the last prime number in the product. For example 2 x 3 = 6. 6 x 5 = 30. 30 x 7 = 210. 210 x 11 = 2310 and so on. These are the pivotal numbers around which the prime numbers (and some false primes which will be naturally eliminated by the Generator Function) are symmetrically generated using only the operations of addition and subtraction.

The first 9 "Magic Numbers"

1	**2**	**2**
2	**3**	**6**
3	**5**	**30**
4	**7**	**210**
5	**11**	**2,310**
6	**13**	**30,030**
7	**17**	**510,510**
8	**19**	**9,699,690**
9	**23**	**223,092,870**

VI. 0 AND 1 – THE BUILDING BLOCKS OF THE PRIMES

With the introduction of the material in the previous chapters you are now ready to start with the initial building blocks of 0 and 1 to generate the prime numbers (all the way to infinity if you have the time). Later we will also see that by using the reverse of this process, you will be able to pick any number and within a few short calculations of addition and subtraction, determine if that number is a prime number or not.

First a word of caution since this material is in a sense bringing the cart before the horse. In a true mathematical presentation one would first define the theorem and then proceed with the application of the theorem. In presenting this for the general public, however, it is easier to present the application before presenting the more complex and abstract equation that I have been calling the "Generator Function".

There are also some advanced topics that many mathematicians would want to discuss including the use of 0 and 1 in the generation of prime numbers. First of all remember that some "false" primes are generated to get to the ultimate goal of the real primes. Secondly, one could collect a room full of mathematicians and get as many opinions regarding esoteric issues such as whether the number 1 constitutes a prime number or not. These are issues beyond the scope of this book and deserve attention but in the correct venues.

Before getting started, for the non-mathematicians reading this book, mathematics is built on basic fundamental building blocks that may seem obvious to some, but which in fact require serious levels of proof. Mathematicians spend lifetimes laboring over these issues.

But for the general public, as a starting point in this book we are assuming that you are familiar with the counting process and the standard definition of prime numbers (a number with just itself and the number 1 as factors). This definition will from this point forward only be used for comparison, as we will actually be re-defining the definition of the prime numbers in terms of addition and subtraction relative to the "magic numbers". The lists of discovered prime numbers will turn out to be the same, but be aware that we will no longer be using "factorization" as a test to see if a number is prime or not. In fact, as we will see later in the book, factorization itself will be redefined in terms of addition and subtraction for *all numbers* (not just primes) relative to the magic numbers.

For many, the beginning will seem a bit trivial and possibly nonsensical, but after all is said and done and the final Generator Function is defined, much of this will make more sense at that time, so please be patient through this initial description. The results of the generation process will be called "prime candidates". Some of these will turn out to be "false" primes and the remaining numbers will end up being the "true" primes as the process proceeds.

Essentially, we will be adding and subtracting 1 and all the "prime candidates" generated in the previous application of the process to the "magic numbers" as they are generated. In the strict mathematical derivation, many other issues are dealt with in stricter terms, but once again this book is meant for the general public (mathematicians may wish to continue with this more elementary description before referring to Appendix I for a more detailed approach). Understand that what I am directing in the following outline is exactly the procedure that is defined by the Generator Function.

Begin with the numbers 0 and 1. First add and subtract 0 from 1. This results in just one number, which is 1. Add and subtract 1 from 1. This results in 0 and 2. Now continue adding 1 to each result to generate the first list of "potential prime candidates". This will result in 2+ 1 = 3, 3 + 1 = 4, 4 + 1 = 5 and so on to infinity. In Number Theory one would say we are generating the Natural Numbers and if you include 0 these would be called The Integers.

Clearly, every prime number (in fact every number) will be contained in the group of numbers that we just generated. There is a fundamental reason for including this step. At some point in the generation of the primes the question will arise, "have we discovered all of the prime numbers?" By starting with all of the numbers and eliminating some and leaving the rest to be "true" primes, this answers that question. In other words, all numbers have been included in the process and either they have been rejected or are included in the final group of primes. Not a stone has been left unturned so to speak.

The elimination process however works very quickly and you will see that before too many iterations of the application of the Generator Function, almost all of the prime numbers all the way out to infinity will already have been identified with only a few percent remaining to be eliminated. This will demonstrate the power of this method relative to the standard factorization method that requires immense amounts of super computing power to locate just one prime number, and after you have identified the one prime number, you have to start all over again. A later chapter will compare the two methods of determining prime numbers.

Before proceeding to determining and using the first "magic number", a side issue with related explanation is necessary to prepare you mentally for wave patterns of numbers that are an essential part of this process.

The following discussion is necessary to develop the importance of "magic numbers" as the key numbers around which the prime numbers are generated.

Establish on a number line the patterns of all multiples of 2, then all multiples of 3, then all multiples of 5, etc. These "repetition patterns" collectively form "repetition groups" that eliminate numbers on the number line from being considered as prime, using the traditional "factorization" definition of prime numbers.

To generate the "repetition pattern" of 2, we would have 2, 4, 6, 8, 10, 12, 14, 16, etc. out to infinity. This group of numbers we call a "repetition group" and denote it with the symbol {2}. Now to generate the "repetition pattern" of 3, we would have 3, 6, 9, 12, 15, 18, etc. again moving out to infinity. Denote the "repetition group" by the symbol {3}. Note that some of the members of the "repetition group" {3} are also members of the "repetition group" {2} (e.g. 6, 12, 18 are members of both "repetition groups". Combining these two groups creates a single repetition group {2,3} and has the following members; 2, 3, 4, 6, 8, 9, 10, 12, 14, 15, 16, 18, 20, 21, 22, 24, 26, 27,28, 30 etc. This "repetition group" has a repetition rate of 6. Notice that the spacing between the numbers repeats as follows:

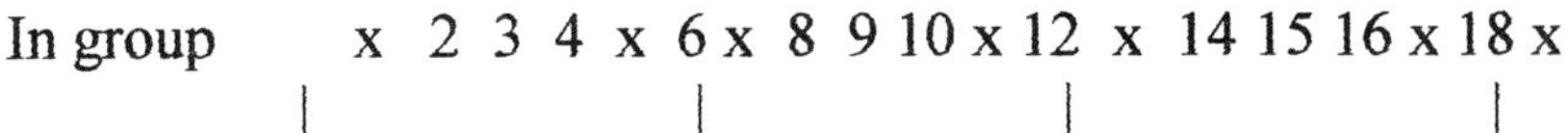

As an exercise carry this list out to about 100.

Later I will use the analogy of a comb with some teeth missing where the x's represent the numbers that are missing. The pattern generated by the "repetition group {2,3} has a "repetition length" of 6.

Now generate the "repetition group" based on 5. It contains 5, 10, 15, 20, 25, 30, 35, 40, 45, 50, 55, etc. Note that the only numbers not already included in the {2,3} "repetition group" will be 25, 35, 55, etc. From a traditional view point, this is because they have factors of 5, 7 and 11 (numbers not contained in the patterns generated in the "repetition group" of {2,3}. As an exercise list the members of the group {2,3,5} and convince yourself that the "repetition length" of this group is 30, that is, the repeating patterns converge on the numbers 30, 60, 90, etc. As above, put x's in the blank spaces that represent the missing numbers in the pattern. Observe the pattern.

In selecting the next number to generate a "repetition pattern", one need only select already identified prime numbers since, for example, all multiples of 4 have already been eliminated by the "repetition pattern" based on 2. This entire process could be used as a method for determining prime numbers, but using this method would not reveal the true and subtle nature of primes. This exercise has only been used to prepare you to use similar groups in the generation of the prime numbers relative to the magic numbers.

What is important from this discussion of "repetition patterns" is that one can define "repetition groups" R_{gn} (where n is the nth sequence "repetition group") based on a sequence of primes and their respective "repetition patterns". The first group $R_{g1} = \{2\}$ is made up of the number 2 and its "repetition pattern". Defining the second and subsequent sequences, $R_{g2} = \{2,3\}$ is made up of 2 and 3 and their combined "repetition patterns", followed by $R_{g3} = \{2,3,5\}$ and next $R_{g4} = \{2,3,5,7\}$, etc. with their combined "repetition patterns", etc. Using {2,3}, for example, this "repetition group" has a "repetition group length" of 6 and repeats the same pattern indefinitely. The "repetition group" $R_{g3} = \{2,3,5\}$ has a "repetition group length" of 30 and again repeats indefinitely, with the {2,3} "repetition group" of length 6 being a subset of this larger "repetition group". Another way of looking at this is that the "repetition group" {2,3,5}, for example, converges on the numbers 30, 60, 90, etc. The "repetition group lengths" are 2, 6 (e.g. 2 x 3), 30 (e.g. 2 x 3 x 5), 210 (e.g. 2 x 3 x 5 x 7), 2310 (e.g. 2 x 3 x 5 x 7 x 11), etc. These are the "magic numbers listed in the previous chapter. They are magical for many reasons that you will learn in upcoming discussions. They are the pivotal numbers needed to generate the prime numbers.

Just one more note on the above patterns and "repetition groups". These will actually be used later in the process, but we will show that by not including these in the Generator Function, that we will in fact in a very organized way eliminate numbers that we already know will never be prime numbers. The trick of course is to find the prime numbers that are the initial numbers of these groups ... these initial numbers are the "true" primes. Once we have identified a "true" prime by addition and subtraction of groups of numbers based on the "magic numbers", then those newly identified primes will have their own "repetition groups" generated and we will no longer deal with any of the multiples of these numbers ever again. You may be starting to see why this method of determining prime numbers is so efficient when compared to the brute force method used by super computers. We directly calculate the prime numbers (not just one at a time but in groups) and once we have these we then eliminate all the other numbers that are in the "repetition group" of that newly discovered prime. We then use those newly calculated primes to generate more and the process continues to infinity.

Also note that we have introduced the fundamental concept of "waves of numbers" and "repetition lengths" which we will rename later as "wavelengths". Yes, the primes are generated in waves ... these are infinite waves that repeat out to infinity. Thus you find repeating patterns in the prime numbers. That is why we see reoccurring prime patterns that arise and then fade away as existing waves of prime numbers are modified by future waves that are generated as we proceed. Mathematicians and physicists who study wave phenomenon will be interested in this amazing property of the prime numbers. It is small wonder that physicists see correlations between the prime numbers and Quantum Mechanics, which of course describes the wave-like nature of the sub-atomic world.

Now returning to the "prime candidate" 2 and continuing with the direct calculation of the prime numbers. We saw that the first generation above actually produced all numbers and there were good reasons for doing this (to guarantee that we have considered all numbers in the process of searching for the primes, and have left no stones unturned). This is why it is important to start with just the numbers 0 and 1. It is also of fundamental significance in the very nature of the Generator Function, that ALL of the prime numbers have their ancestry in just the numbers 0 and 1.

The next generation process will reduce the number of "prime candidates" significantly. At this point we will claim that 2 is the first prime number. Multiply 2 x 1 to obtain the first "magic number". The first true magic number is 2.

Next, form the "repetition group" based on the prime number 2. These numbers would normally be called "multiples of 2" but in this definition they are called the members of the "repetition group" of the first "magic number" which is 2. This is the only case is which a "magic number" is also equal to the prime number but for clarity and consistency, it is important to distinguish between the two (this will never happen again so bare with this issue as we get started in this process).

The process of adding 1 and all other known prime numbers (a null set with no elements) to 2 forms the next set of "prime candidates" which constitute the list of "prime candidates" of the "magic number" 2.

$$2 \pm 1 = 3,1 \text{ (3 and 1 are solutions)}$$

Since there are no other "prime candidates" identified at this time, 3 is the only new "prime candidate" discovered using the "magic number" 2. Since 3 has now been identified as a "prime candidate", it can be added to 2 to generate more "prime candidates". 2 + 3 = 5, so 5 can now be added to 2 to give 7, etc. to infinity. This process continues, but almost immediately it seems to break down, as 2 + 7 = 9, which we know is not a prime. This is where the next sequence will display how "eliminator rules" work. We will see that 9 is to be eliminated when generating the next group of "prime candidates" associated with the next "magic number". In future "generations", this elimination will not occur with the next generation of "prime candidates" but will require other future generations to eliminate the "non-prime" from the current list of "prime candidates".

Once again every prime number is contained within this remaining group (which is made up of all the odd numbers out to infinity). Note that we have already eliminated a healthy population of numbers to get to this group. In fact, we have eliminated 50% of all the numbers out to infinity, a healthy amount for the first pass of the process. To the trained mathematician, this at first might look like a simple elimination similar to using multiplication. This similarity will be short lived.

Now we must apply a test to determine which numbers in our list have been solidified to become true primes. Once again I will put the cart before the horse and state that every "prime candidate" in our list that is less than $2^2 = 4$ will be set in stone as a prime number. For clarity, the 2 used here is the highest prime factor of the 'magic number" 2. Once again we have to differentiate between the "prime number" 2 and the "magic number" 2. This complication will not occur again as we proceed into larger groups of generating the primes.

The reasoning and proof for using $2^2 = 4$ is part of more complex mathematical treatments and actually you can see that many of the numbers in our most recent list are in fact primes. The method for determining which numbers in the list are prime is a complicated subject, but the rule just stated works every time and is simple. It is sufficient to continue on to infinity with the process being followed here. Thank you for your patience in wading through the initial first pass of this process, now you will begin to see the power and elegance of calculating primes based on the Generator Function and using just addition and subtraction. With each successive pass of the process we will begin generating more and more prime numbers in groups.

Since 3 is less than $2^2 = 4$, then 3 is now a "true" prime. Our list of primes now includes 2 and 3. So let's find the next "magic number" by multiplying 2 and 3 together to get 6. Begin the process again. Add and subtract 1 from 6. From here on we will use the notation

$$6 \pm \quad 1 = 5, 7$$

This indicates that the two "symmetrical" numbers 5 and 7 have been generated around the magic number 6. At this point 5 and 7 are just "prime candidates" (we already know from experience that they will be "true" primes, but watch how this is accomplished). Now add and subtract the entire list of numbers formed above to 6, except for the numbers that are factors of 6 (2 and 3) and their repetition groups. The reason for this will become clear later in the book, it is an essential part of the Generator Function definition that will be presented fully in detail later in the book. Remember that I am leading you through the method before detailing the mathematical equation that describes this process. The results of the present calculations around the magic number 6 are listed below.

$$
\begin{aligned}
6 \pm \quad 1 &= 7, 5 \\
5 &= 11, 1 \\
7 &= 13, \text{n/a} \\
11 &= 17, \text{n/a} \\
13 &= 19, \text{n/a} \\
17 &= 23, \text{n/a} \\
19 &= 25, \text{n/a} \\
23 &= 29, \text{n/a} \\
25 &= 31, \text{n/a} \\
29 &= 35, \text{n/a} \\
31 &= 37, \text{n/a} \\
35 &= 41, \text{n/a} \\
37 &= 43, \text{n/a} \\
41 &= 47, \text{n/a} \\
43 &= 49, \text{n/a} \\
&\ \ldots
\end{aligned}
$$

The 3 dots at the bottom of this list show that this process continues to infinity. The notation n/a indicates that by subtracting the number from 6 you would get a negative number and therefore is "non-applicable" as we are only dealing with the positive numbers in the case of the primes.

Next apply the rule to solidify the "true" primes from this list. The highest "prime factor" of the magic number 6 was 3, so $3^2 = 9$ is the dividing line. Thus 5 and 7 move from "prime candidate" status to confirmed "true" primes. Our complete list of primes now includes 2, 3, 5 and 7. Once again you will see many real primes in the list above and more powerful criteria could claim them at this point, but we are going to use the basic method already in use. All of the rest of the numbers in the list are considered "prime candidates" and on the next round we will modify it further and find more "true" primes and also further reduce the number of "prime candidates" that will remain for the next operation.

At this point a number of observations are in order. First of all we are both adding and subtracting the entire list of "prime candidates" to the magic numbers, except for the members of the "repetition group" {2,3}. Notice also that there is a pattern that emerges that repeats and repeats and repeats. With the magic number 2 our repetition pattern had "length" of 2. With magic number 6 the repetition rate was of "length" 6. This means that we are truly generating a literal "wave of numbers" both to the left of the magic number and also to the right.

Imagine this pattern of "wavelength" equal to the magic number to be like a comb with certain teeth missing (the eliminated numbers). The teeth that remain are the "prime candidates". Now make up millions and millions of these combs ... all exactly the same. Put them end-to-end out to infinity in the positive direction starting at the magic number. Now ... turn ONE of your combs around and place it to the left of the magic number (it just takes one wave cycle or comb to reach zero, so that is where the negative directed wave ends).

This is your wave ... a wave of "prime candidates" and it propagates in both directions starting from the "magic number". But something very exciting is happening to the left or negative side of the magic number in your wave. Some of the numbers are being solidified to become "true" and permanent members of our prime number list. Not just one prime is generated, but an entire group of primes. Each group is generated from the previous group ... thus the name "Generator Function". This process has a somewhat complicated mathematical formula, but this is what it does. It operates on already existing groups of "prime candidates", reducing the group each time and solidifying some of the members into permanent prime status.

With this in mind let's generate the next group and make some more observations. In the list above generated around the magic number 6, note that the only numbers there that are not primes (check your prime number table if necessary) are 25, 35 and 49. Of course if you continued to infinity there would be lots of "false" primes in the list. But watch what happens to 25, 35 and 49 in the next iterations of the process. They will be necessary to generate certain "true" prime numbers and when their usefulness has ceased, they will be naturally eliminated from the future lists by this process. They will hold their place and balance the symmetry of the process (the symmetry generated by adding and subtracting the prior list of "prime candidates" to the next magic number). As the tables get larger and we get into the very large numbers, it will be impossible to tell which are true and false primes, and we will simply have to let the process take its course.

What a subtle and interesting result. One must generate false results that are needed to generate true results ... and these false results are then eliminated by the very same process after their usefulness is complete. Possibly now you are beginning to see why centuries of mathematicians worked around this problem without seeing through this thin but ever so illusive curtain.

Let us move on to the next magic number. 5 is the next prime number in our growing list so we multiply the prior magic number by 5, thus 6 x 5 = 30 is the next magic number. You already have had some experience with this magic number in the previous chapter.

Add and subtract 1 and all the members of the prior list to 30, except of course all the members of the "repetition group" {2,3,5}. Note that the list we will use starts at 7 and does not include 2, 3 or 5 (the prime factors of 30) or any members of that "repetition group". When the Generator Function is defined in a later chapter, you will readily understand certain terms that perform this function in the mathematical equation.

There is a very good reason for this. This is another subtle aspect of this process and relates to a fundamental law of algebra. Any number or multiple of a number added to 30, which is also a factor of 30, will result in non-prime number. There is another aspect to this process that may not meet the eye at this point. The question is, "how do we know we are in fact creating all of the primes?" The mathematical proof of this in fact lies in the last statement and technically is beyond the scope of this book, but in simple terms, we started with ALL of the numbers. Either they have been eliminated by the generator function or they are left remaining as "true" primes.

Watch what happens to our good friends 25, 35 and 49 in this next iteration of the Generator Function as the process now works on the last group of "prime candidates" that were generated. They will be used to generate a new list of "prime candidates" in the positive direction from the magic number 30, but to the left (e.g. in the negative direction from 30) they are eliminated from the list where the solidifying primes are congregating. 49 on the other hand is used to generate the future true prime 79 (30 + 49 = 79) and is still maintaining its place in the symmetrical structure around the magic number 30. But watch what happens to 49 in the next round. Interestingly 30 + 47 = 77 which is not a prime number. Watch these numbers on the subsequent rounds.

Following our procedure, the next list of "prime candidates" is contained in the following list.

$$
\begin{aligned}
30 \pm \quad 1 &= 31{,}29 \\
7 &= 37{,}23 \\
11 &= 41{,}19 \\
13 &= 43{,}17 \\
17 &= 47{,}13 \\
19 &= 49{,}11 \\
23 &= 53,\ 7 \\
29 &= 59,\ 1 \\
31 &= 61,\ \text{n/a} \\
37 &= 67,\ \text{n/a} \\
41 &= 71,\ \text{n/a} \\
43 &= 73,\ \text{n/a} \\
47 &= 77,\ \text{n/a} \\
49 &= 79,\ \text{n/a} \\
&\ldots
\end{aligned}
$$

Finally, using the rule to solidify the primes to the negative side of 30, using our rule ... since 5 is the largest prime factor of 30, then $5^2 =$ 25 gives the upper limit for "true" primes. Therefore 11, 13, 17, 19 and 23 join the growing list of prime numbers. In the first pass of this process we generated just one true prime number, now we are generating 5 in one pass. The power of this method is hopefully becoming clear.

Remember that we will need an ongoing supply of new prime numbers to keep this process going. Since each group generates more true primes than were generated in the prior group, then this process will continue to infinity, with a rich supply to provide us with more and more "magic numbers" to keep the process going. We are directly calculating the prime numbers using only addition and subtraction, and each time we apply the process, the list of eligible primes becomes smaller as we look out on the "wave" of numbers that extend out to infinity. In a later chapter we will compare this process to the traditional process of factorization in determining primes, which grows in complexity rather than decreases as you discover more prime numbers.

One more point shows the intricate and subtle nature of the prime number system when viewed from this perspective. When generating the prime numbers to the left of the current magic number, the prime numbers agree exactly with the prime numbers generated in the positive direction from all prior magic numbers. As shown in our original examples in the last chapter, one can also start with any multiple of any of the magic numbers and find similar patterns in the primes around that number. These uncanny and amazing relationships show that the primes are an amazingly complex and inter-related set of numbers. ***They are anything but "random".***

Going to the next magic number, 7 is the next prime after 5 so 30 x 7 = 210 gives the next magic number. Following the same process calculate the next set or prime candidates and determine which primes are in fact "true" primes and add them to the list. The process here gets more complex as more and more "false" primes begin to turn up that will not be eliminated until 2 or 3 or even more iterations of the application of this process. At this point if I continued, many average people would become confused and lost in the complexities of the process. If you are mathematically adept and have the time, you are free now to continue in this process.

But for the less mathematically inclined I will start to talk about patterns on the prime number table, and actually this is where the fun starts. Remember that I said earlier in the book to take a good long look at the prime number table, because it would never look the same to you again. So look at it now and start looking for the inter-related patterns around the various magic numbers and the multiples of these numbers.

There are some very interesting issues to deal with here. If for example you are generating the "prime candidates" around the magic number 210. You will have a large list of symmetrical numbers around 210 after adding and subtracting 1 from 210 and all the members of the prior list of "prime candidates" (of course after eliminating from this list the members of the "repetition group" {2,3,5} of the previous iteration of the magic number 30 as the process requires). The numbers 25, 35, 49 and 77 that I have been talking about all meet their end in the process defined here when we arrive at the "magic number" that contains 5 or 7 or 11 in their construction. At that point when we SUBTRACT the appropriate "prime candidate" numbers from the magic numbers, these numbers are no longer found in the mix. For example, with the magic number 30, 25 and 35 are eliminated, whereas 49 is eliminated with the magic number 210 as we subtract the appropriate "prime candidates" that belong the group related to 210. Remember that all of this happens as a natural by-product of the process that has a relatively simple equation that will be explained term by term in the following chapter and which performs exactly the process we have been following here to directly calculate the prime numbers. Another chapter later in the book will relate this to basic mathematical properties of addition, while giving a completely new definition of the process we traditionally call "factorization".

There is one point that I learned as I studied physics and mathematics over my career both as a student and as a professional. You rarely if ever fully understand a concept the first time you read it. You usually have to read and think about new concepts many times before all of it makes sense and you start to associate the many aspects of a new topic. So if at this point things make little or no sense and you feel that you really are not getting the "big picture", you are most likely not alone. In fact, you most probably are in very good company. So read the first time for the overview. Read the second time for basic understanding of the flow of material. Read a third time to start to understand. And read again for deeper understanding.

****

Sufficient information has now been presented to allow readers to continue and generate all future primes and groups of "prime candidates". A few further examples will be given to clarify some of the statements made throughout this paper. With this further discussion we will start to migrate into using more mathematically correct symbols to prepare you for the next chapter, which finally defines in detail "The Generator Function". You are now ready to see and understand it and to make sense out of the various terms that otherwise would look like so many Greek symbols. I will start using the term "sequential prime product" (using the symbol $S_{pp\,n}$) in place of "magic number" and the term "umbrella group" (using the symbol U_{gn}) to represent the list or "prime candidates" from the previous iteration of the process with the appropriate "repetition group" members removed. If you are not a mathematician and these symbols bother you, then this would be a good point to leave this chapter and go on to Chapter VII. But before you do, let me leave you with a parting thought.

Are you beginning to see why possibly ages of extremely talented mathematicians have worked around this problem and stared it in the face their entire lives without solving the entire problem? There are many subtle and illusive aspects to this problem as you now can see.

Are the prime numbers random? You don't have to take my word for it anymore. You can now determine this for yourself.

For those that are remaining in this chapter, the following are some further thoughts on the process of calculating the prime numbers.

In the "sequential prime product" based on the "repetition group" of $R_{g4} = \{2,3,5,7\}$, $S_{pp\,4} = 210$ is used in the generation of its "umbrella group" of "prime candidates". 210 + 1 = 211, which is prime. However 210 – 1 = 209, which is not prime, and will be eliminated by the next "umbrella group" which will include 11 in its "repetition group". 11 is the first prime of $U_{g(4-1)} = U_{g3}$ and not a member of R_{g4}, so 210 - 11 = 199 which is a true prime, however 210 + 11 = 221 which is a non-prime that will not be "eliminated" with the next "umbrella group", but will in fact have to wait until the second future "umbrella group" which will include 13 in its "repetition group". As already noted, the process of elimination becomes more complex as future "umbrella groups" are generated, but the fundamental rules remain the same.

Remember, the non-prime "prime candidates" in an "umbrella group" are essential to the complete generation of all primes.

The primes added to $S_{pp\,4} = 210$ which will eventually prove to be true primes are: 1, 13, 17, 19, 23, 29, 31, 41, 47, 53, 59, 61, 67, 71, 73, 83, 97 and so on. Likewise, the primes subtracted from 210 that will eventually prove to be true primes are: 11, 13, 17, 19, 29, 31 and so on. It is a relatively simple exercise to complete this list by first establishing the entire "umbrella group" U_{g4} of the "sequential prime product" 210 and its subsequent modification by future "umbrella groups" until the above mentioned "stabilized" list is achieved.

Going to the next "sequential prime product" $S_{pp\,5} = 2310$ the prime numbers which are added to generate the true primes of this "umbrella group" turn out to be 1, 23, 29, 31, 37, 41, 47, 61, 67, 71, 73, 79, 83 and so on. The prime numbers which are subtracted from 2310 to generate the true primes in the negative direction are 1, 13, 17,

23, 29, 37, 41, 43, 59, 67, 71, 73, 79, 87, and so on. . It is a relatively simple exercise to complete this list by first establishing the entire "umbrella group" U_{g5} of the "sequential prime product" 2310 and its subsequent modification by future "umbrella groups" until the above mentioned stabilized list is achieved.

Any prime number can be calculated from any "sequential prime product" $S_{pp\,n}$. For example, select 1427, which is a prime number. From 210 add 1217, which is a prime number (210 + 1217 = 1427). From 2310 subtract 883, which is a prime number (2310 – 883 = 1427). Or one can add prime numbers as in the example 2310 + 1051 = 3361; all of these being prime numbers.

One must be careful since there are some values of the differences between prime numbers and "sequential prime products", which will not result in true prime numbers. These are due to the fact that the non-prime differences are "prime candidates" of the "umbrella group" U_{gn} associated with the assocated "sequential prime product" $S_{pp\,n}$. As an example, 331 is a prime number. 331- 210 = 121, but 121 is not a prime number. This is explained by the present method of viewing primes by the fact that 121 is in the "umbrella group" $S_{pp\,4}$ = 210. 121 will be "eliminated" from use in generating prime numbers in the "umbrella group" U_{g5}. So until that point any number that is a member of a higher order "repetition group" such as 11 (or other higher order primes for that matter) may be found in the prime tables. Likewise, the "sequential prime product" $S_{pp\,5}$ = 2310 finds a similar example. The number 1409 is prime. 2310 – 1409 = 901 which is not prime (e.g. the difference between $S_{pp\,5}$ and the prime number 1409 is not prime). But 901 will not be eliminated until the "umbrella group" U_{g7} associated with the "sequential prime product" $S_{pp\,7}$ = 510,510 and its related "repetition group" R_{g7} containing the prime number 17. The operator guarantees that every false prime will identified and only the true and complete list of primes will remain up to the criteria set previously (for all numbers less than the square of the largest member of the "repetition group").

There are strange patterns that emerge such as the sequence of primes starting with 907: 907, 911, 919, 929, 937, 941, 947, 953, 967 and 971. These seemingly follow the pattern of primes 7, 11, 13, 17, 19, 23, 29, etc (but with some numbers missing). This is what was noted earlier. These patterns seem to arise and then fade as new patterns emerge. Likewise the same pattern emerges again at 1,511 and continues in the 1,600's but again fades away: 1511, 1523, 1531, 1543, 1549, 1553, 1559, 1567, 1571, 1579, 1583, 1597, 1601, 1607, 1609, 1613 and 1619. At this point the "wave" begins to fade and new patterns emerge. Note that 900 and 1500 are both multiples of $S_{pp\,3}$ = 30. Similar patterns will be obvious around all multiples of the infinite list of "sequential prime products", all of which give rise to waves with wavelength of the same values.

In spite of these unusual patterns, the entire spectrum of primes can be calculated given the methods described in this paper and understood in terms of the periodic "waves" of "prime candidates" and ultimately "stabilized" prime numbers generated in the succession of "umbrella groups" with repetition "wave lengths" equal to the "sequential prime products" = $S_{pp\,n}$. Additionally, one can be assured that all of the primes can be identified without exception, that is, the generated group of prime numbers is complete.

Associating prime numbers follows from this discussion. For example, adding and subtracting the prime number 7 to any multiple of 30 will generate certain primes. The following table shows related primes in this family of primes based on wavelength 30.

$30 \pm 7 =$ 23,30
$60 \pm 7 =$ 53,67
$90 \pm 7 =$ 83,97
$120 \pm 7 =$ 113,127

However 150 - 7 gives 143, which will be eliminated by the "repetition group" containing 11. 150 + 7 = 157 will be discovered to be a true prime and will not be eliminated by any future application of the operator. One can also view all of these numbers from the vantage point of $S_{pp\,4}$ = 210 (or any future "sequential prime product" $S_{pp\,n}$). For example 210 – 187 = 23 where 187 is not a true prime but is in the "umbrella group" of 210 (e.g. $S_{pp\,4}$ = 210 still has 11 in its "umbrella group"; 187 will not be eliminated until the operator works on the "umbrella group" with "repetition group" R_{g5} = {2,3,5,7,11}). 210 – 173 = 37 where both 173 and 37 are true primes. Even though 210 ± 7 will not produce a prime (since 7 is in the "repetition group" of 210), some future products of 30 (e.g. n x 30 ± 7) which are greater than 270 will produce primes such as 270 + 7 = 277. However there will never be another prime number of the form n x 210 $\pm$ 7 (the wave based on wavelength 210, n = 1,2,3....).

The point is that the placement of prime numbers in the prime number table is not random and the prime numbers can now be understood relative to any of the "sequential prime products" in terms of waves and relationships to all the other prime numbers originating with the numbers 0 and 1. The essential basis of this is the operator defined in the Alternative Definition of Prime Numbers. The concept of taking a primitive set of numbers such as 0 and 1 and generating the entire set of prime numbers is a tribute to this powerful mathematical tool and holds promise for future applications with more complex building blocks in conjunction with more complex generator function operators. This has applications in the realms of genetics, quantum mechanics, and complete analytical solutions to N-Body problems such as in celestial mechanics as well as muli-dimensional topology. The solution to the Riemann Hypothesis may be closer using this as a basis for understanding the true nature of the prime numbers. The most illusive and long-standing mathematical problem in both Mathematics and Physics, the method of directly calculating the prime numbers, is found in this solution.

VII. THE GENERATOR FUNCTION

After reading and calculating through the exercises in the earlier chapters you are now ready for the true mathematical formulation of the "Generator Function".

Note in the very definition the symmetry of the true and false primes is guaranteed because you are adding and subtracting the exact same numbers from the magic numbers, which are actually called "sequential prime products". First look at the name and what it means. "Sequential" means all in sequence or in an orderly succession. "Prime" indicates that we are using just "true" prime numbers and the "product" means they are multiplied together to get the result ... thus the term "sequential prime product". These are the magic numbers around which the prime numbers are formed.

The true theoretical basis for this is actually quite simple and is based on a fundamental property of algebra called the "distributive property" that is used to factor out a member of a sum of numbers. For example 21 + 15 = 36. But one could equally write this expression as 3 x (7 + 5) = 3 x 12 = 36. We have taken out the factor 3 and added the remaining factors 7 and 5 to get 12 and then multiplied the removed number 3 to 12 to get the same result. This is only possible when the two numbers (e.g. 21 and 15) have the same factors.

If we create a number that contains all of the prime numbers up to a certain point (the sequential prime product), then any other test number that has any one or more of these same prime numbers as a factor should allow use of the distributive property to obtain a result. However, if our test number DOES NOT have any of the factors in the list, then it is called "relatively prime" and we have generated a "prime candidate". If the prime candidate stands the test of being compared and compared to repeated "sequential prime candidates" up to the square of this test number, then it is guaranteed to be prime. This is the basis of the process used in the generation of the prime numbers presented in this book.

Now revisit the example of 21 + 15 = 36. Let us take the difference between the 2 numbers 21 and 15. 21 – 15 = 6. You see that 6 has 3 as a factor. We just "factored" all of the numbers in this example by using subtraction. We could reverse this process and add 6 to 15 to determine that 21 is in fact a number containing factors of 3. Are you starting to see that we do not need the process of factorization to determine what factors are contained in a number?

So if we take only those very pure numbers that only contain a single one of every prime number up to a given value (the sequential prime products), we can determine all of the numbers that contain these primes as factors also by just using addition and subtraction based on these "magic numbers". Likewise, we can also determine all the numbers that DO NOT have any of these factors in common, and these are what we have been calling the "prime candidates". As already noted, at a certain point, no matter how many more "sequential prime products" we generate to test our number, at a certain point the future "sequential prime products" will no longer have a change to "eliminate" our test number because they are simply too large. If our test number survives the gauntlet, then it will become a "true" prime and be added to the list of primes waiting to generate its own "sequential prime product" and be used to test future large numbers.

This in simple terms is the process used to determine all of the prime numbers using only addition and subtraction. One of my challenges was to create single mathematical equation that reflected all of the operations needed to encompass this process, just as for example, a differential equation or wave equation would describe a physical situation in nature.

First you will see the Generator Function and then I will break it apart piece by piece so you understand what it says and how it "operates" on successive groups of "prime candidates" that were generated by the prior pass of the Generator Function. The process starts with just 0 and 1 and ends with the complete list of prime numbers to infinity. This is a tall order for a single equation and shows that possibly the mathematical method is as significant as the solution of the age-old prime number problem itself. You will see that inherent in the equation is the properties of symmetry (where "prime candidates" are symmetrically generated around the "sequential prime products") and "reciprocity" (where the primes generated in the negative direction from a given "sequential prime product" are exactly the same as those generated in the positive direction from ANY other prior "sequential prime product"). These are indeed amazing properties, especially for a group of numbers that previously were thought to have no patterns or rhymes or reasons at all.

The application of this mathematical technique to other problems will be most interesting, and already my personal research is leading down the road to solutions of other unsolved problems in both mathematics and physics. In the future it will be interesting to see where the concept of Generator Functions will take us.

With this in mind, the Generator Function used to calculate the prime numbers starting with just the numbers 0 and 1 is defined as follows.

********** THE GENERATOR FUNCTION ***********

The notational representation for an "umbrella group" is as follows (the symbols ε and $\underline{\varepsilon}$ represent "are an element of" and "are not an element of" respectively):

$$U_{gn} = \Pi_1^n p_i \pm 1 \text{ and } \Pi_1^n p_i \pm p_{cm}$$

$$p_{cm} \varepsilon U_{g(n-1)} \text{ and } p_{cm} \underline{\varepsilon} R_{gn} \text{ (for all } p_{cm} > 0)$$

****

Each term will now be separated out and examined individually.

In this definition the "sequential prime product" is written by the following expression.

$$S_{pp\,n} = \Pi_1^n p_i = \text{"repetition group length"}$$

(Where p_i is the ith prime number of the nth "sequential prime product" $S_{pp\,n}$ and the Greek letter Π represents the operation of multiplying all terms defined after it).

For example ... $S_{pp\,5} = \Pi_1^5 p_i = 2 \times 3 \times 5 \times 7 \times 11 = 2{,}310$

This term contains the first 5 prime numbers as factors

$\mathbf{U_{gn}}$ is called the "umbrella group" and is formed by adding and subtracting 1 from the current "sequential prime product". The subscript "g" stands for "group" and the letter "n" indicates that this is the "nth" umbrella group associated with the "nth" sequential prime product". To translate this for non-mathematical readers, we started with the first "sequential prime product" $\mathbf{S_{pp\,n} = 2}$ and so we added and subtracted 1 from 2 as you recall.

"and" is a conjunction that indicates that you must not only add and subtract 1 from the "sequential prime product" but you also must add and subtract all prime candidates $\mathbf{p_{cm}}$ from it also. "and" is a term used in mathematical logic indicating you have to include both expressions on the two sides of the word "and" (if the intention would have been to perform one or the other of the operations, we would have used the word "or", but the intention was to include both calculations, and thus the use of the word "and"). The subscript "c" stands for Prime candidate when used with the "p" and the letter "m" indicates that we are using numerous prime candidates. Note that the entire next line is dedicated to defining the values that we will use for the variable $\mathbf{p_{cm}}$.

The entire second line of the Generator Function gives what are commonly called "boundary conditions" in mathematics. It defines and limits the values that the variable $\mathbf{p_{cm}}$ will take as we calculate the values of this equation.

Now to examine the second line

$\mathbf{p_{cm}}\ \varepsilon\ \mathbf{U_{g(n-1)}}$ says that $\mathbf{p_{cm}}$ will have as its values all the members of the prior umbrella group (the "(n – 1)th" umbrella group) that we generated on the last iteration of the process ... remember that this is what we did in the examples in the prior chapter.

"and" ... here is the "and" word again which indicates that the values of $\mathbf{p_{cm}}$ will take on the values already noted above "and" will

NOT contain any of the members of the current "repetition group" ... and is demoted by the following expression ... $\mathbf{p_{cm}} \underline{\varepsilon}\ \mathbf{R_{gn}}$.

Last but not least, the expression **(for all $\mathbf{p_{cm}} > 0$)** indicates that we are dealing with only values of prime numbers greater than 0 (we are not including 0 or any negative numbers, only 1, 2, 3 and so on).

So part of my work after discovering the method of calculating the prime numbers using only addition and subtraction and beginning with just the numbers 0 and 1, was to write this in the form of a universal mathematical expression which included the boundary conditions and method of using the equation. The Generator Function generates an "umbrella group" using an already existing "umbrella group". Thus by its very nature it is repetitive and does exactly what the previous chapters were doing as we started the process of calculating prime numbers.

Notice that the "sequential prime products" (what were originally called the "magic numbers") are defined in the equation. By the very nature of the equation, the members of the "umbrella group" (what we had been calling the "prime candidates") create symmetry in the next "umbrella group" around the "sequential prime product" because we both add and subtract every term from this "magic number". By the very nature of the equation it guarantees the property of symmetry in generating the prime numbers.

And last but not least, the process provides the process for solidifying the "true" primes and eliminating the "false" primes when their useful lives have come to an end. Remember that the "false" primes are needed to maintain the symmetry of the process in conjunction with the "true" primes and are additionally needed to generate the "true" primes. Without them our list of prime numbers would break down and be woefully incomplete. After their usefulness has ended, the "false" primes are eliminated naturally from the process. The primes are solidified to the left or negative side of the "sequential prime product", and to the right or positive side, they generate infinite waves of "prime candidates" that will be tested until they are either eliminated or become "true" primes. With each successive iteration or application of the Generator Function, we generate increasingly more prime numbers in groups, while the number of eligible "prime candidates" becomes smaller. Within a very few iterations of this process, we have literally eliminated the vast majority of numbers from being considered as "prime candidates".

Notice that each successive "sequential prime product" is very much larger than the prior "sequential prime product" (see the list on page 11 of this book. The gaps between these numbers grow very rapidly very quickly.

Before ending this discussion a brief explanation of "factoring" will help solidify a concept that is central to this work. That is, given any number, how can you determine if it is a prime or not? This is of utmost importance. The procedure involves the process of factorization already discussed but additionally it requires keeping a record of the "umbrella group" associated with each "sequential prime product" that has already been calculated and used to generate higher order "umbrella groups".

To determine if a number is prime or not, select the nearest "sequential prime product" and take their difference. If the difference IS NOT a member of the "umbrella group" associated with the "sequential prime product", then the number IS NOT PRIME. If the difference is a member of the "umbrella group" of the "sequential prime product" then it may or may not be prime. You may have chosen a "false" prime that needs to be tested against higher and higher order "sequential prime products" until you have exhausted all "sequential prime products" which are less than the square of the number being tested. If you think about what you are doing, you will discover that this is extremely logical. The tests are all performed by addition and subtraction, never resorting to "factorization" as in the traditional method of determining whether a number is prime or not.

Another way of determining if a number may be prime or not, especially a really large number, is to follow the wave patterns generated in the umbrella groups and see if the number in fact falls on a tooth of your "comb" for the given wave up to a given "sequential prime product". This can be effective since a very high percentage of non-primes and false primes are eliminated very early in the process of applying the Generator Function.

VIII. PROPERTIES OF PRIME NUMBERS – THE UNEXPECTED

Every time I attempt to illustrate the many properties of the prime numbers to people who previously thought they were just random numbers, I try to list the many properties of the primes. But inevitably somewhere in the middle of the list I start to get blank stares as if there were just too many properties to put all on one side of one piece of paper.

Symmetry, reciprocity, wave like nature with repeating harmonics, factorization redefined in terms of addition and subtraction, and the ability to take any number, no matter how large, and with a few short calculations determine if it is prime or not. Then if it turns out to be prime, one can fit it into nearly an infinite number of rhythmic families of prime numbers and relate it through these families to the ultimate ancestors of 0 and 1. This is a lot to comprehend in one sitting, even for seasoned mathematicians and physicists. You may well imagine the hair on the back of my neck literally standing up on end and chills running up my spine as these realizations came to me for the first time. It was if some strange dream was occurring and any minute someone would wake me out of a deep sleep to return to "reality".

There are certainly more associations that still lay sleeping and awaiting other researchers to find. But with that in mind, a few more observations are in order.

There are numerous other topics that will be mentioned briefly that have been developed but which are outside the scope to this book. These include the following.

It will become apparent that "natural" bases of number systems for working with prime numbers will be 2, 6, 10, 30, etc. These will allow various unique views of the prime numbers.

The new Alternative Definition of Prime Numbers has an associated proof for the infinite number of primes. Also, the nature of the "umbrella groups" gives rise to studies of prime numbers using wave analysis, with the ability to program "searches" for various prime patterns and ultimately identify them with minimal computing effort.

The density of primes and theoretical work on proofs such as the number of primes between the squares of numbers may be viewed differently. It now may make more sense to talk about the number of primes between the squares of successive prime numbers rather than the squares of all numbers. The "density of primes" which has been such a central problem throughout the history of the prime numbers and has given rise to "The Riemann Hypothesis" itself, may very well take on a new dimension as researchers start to study it in terms of "density relative to the wavelengths" or "sequential prime products" rather than the linear density referred to in The Riemann Hypothesis and other density equations. The logarithmic equations in a sense do this in their representations, but one can see that the density necessarily is reduced as more primes are discovered, therefore the logarithmic expressions will vary more from reality as one proceeds to larger and larger prime numbers.

The computational efforts to arrive at prime numbers will be very different from the traditional method of factorization. The amount of computation required based on the factorization method has diminishing returns on every aspect of the process. As the potential prime numbers get larger, the process is complicated by the fact that almost every number has to be tested, the number of factors increases as the numbers get larger, the number of primes needed to be included in the test grows, and the density of primes goes down, so there are increasingly more negative result tests. It is a process of diminishing returns. The new method of locating primes shows that the prime lists are stabilized relatively quickly, the candidates are never "tested" but are directly calculated from an ever decreasing list of "prime candidates". Additionally, as already noted, patterns and relationships can be programmed into the search for identification of designated prime associations.

The study of the symmetry of primes is a topic that will take on a new future as the relationships between past and future prime groups are studied. The importance of the "false" primes will take on new importance as they are needed to hold the places to guarantee symmetry in generating the primes, and then are eliminated after their usefulness is completed.

Factorization of any number now can be defined as the difference between any number and any of the "sequential prime products" $S_{pp\,n}$. This is because the "sequential prime products" are based on the prime numbers. For example, the number 81 is subtracted and added to the magic number $S_{pp\,4} = 210$. $210 - 81 = 129$. Both 81 and 129 have 3 as factors. $210 + 81 = 291$. 291 and 81 both have common factors.

Using this method, any number can be tested to be prime or not but with the requirement that one needs to know the members of the "umbrella groups" associated with the "sequential prime product" being used in the test.

Last but not least, the concept of the Generator Function, as applied to the primes, gives a mathematical took that is fundamentally simple, yet is extremely powerful. Who would have thought that one could generate the entire prime number system based on a single simple equation and starting with just 0 and 1? The possible list of other applications, not to mention the understanding of the generation process, may have far reaching effects in other remote areas of research. It contributes to the understanding of the prime numbers in a way in which they have never been viewed before.

There is now a new deeper understanding of the prime numbers that offer solutions to other more complex problems. Many of these are the topics of the other as yet unreleased papers.

IX. FACTORIZATION & THEOREM OF CALCULATING PRIMES

There is a fundamental and formal statement of what is presented in this book. It relates to the new concept that what was once called "factorization" now has a completely new definition and meaning. Furthermore, it relates to one of the essential pieces of the puzzle needed to understand the prime numbers. It uses a combination of basic mathematical principles already known, but there is a unique combination that sets this apart from the individual parts.

In brief, this is based on what we have called the "magic numbers", or in mathematical terms called the "sequential prime products". These are the central points around which the prime numbers are generated and which give them their wave like patterns (the value is in fact equal to the "wave length"). These reference points allow us to locate any number in the prime number table and understand why it is there, as well as to understand why other numbers are not present.

The traditional term "factorization" or "to factor a number" means literally to divide the number by known prime numbers smaller than the given number to determine if the division has a remainder or not. For example 20/2 = 10 therefore 2 and 10 are factors of 20. But 20/3 = 6.66666 which has a remainder of .66666 and therefore 3 is not a factor of 20. This process has been claimed to be the only way of identifying prime numbers and super-computers use essentially this brute force method to determine if a number is prime or not.

The new definition of "factorization" is based on strictly addition and subtraction and the same "magic numbers" given in this book. It is based on the fundamental theorem of arithmetic called the "distributive property" which states that $ab + ac = a(b + c)$.

If one of the numbers (e.g. the product ab) is a magic number, then one can determine if the second number is "relatively prime" by taking the difference between the two numbers and determining if the result has any factors of the magic number. In the new world of primes, this is accomplished by comparing the result with the known table of "prime candidates" associated with the magic number. Division is NEVER used, so factorization takes on an entirely new meaning. The brute force method of calculating factors is replaced by a single subtraction from the nearest "magic number" and that number is compared to the "umbrella group" otherwise known as the set of "prime candidates" of the "magic number". That is all.

The following mathematical expression states this fact:

Given $S_{pp\,n} - A = p_1 \times p_2 \times \ldots p_j \times \ldots p_{n-1} \times p_n - A = B$

If A has p_j as a factor, then B will necessarily have p_j as a factor.

One can also determine if A has a factor p_j if B has it as a factor.

If B does not contain any of the factors in the list p_1 to p_n, then it is said to be relatively prime to $S_{pp\,n}$ and either is a "true" prime or it has factors that will be tested and identified as a non-prime with an expanded list of prime numbers in a future application of the "Generator Function" in our methodical process of generating the prime numbers. But there is a limit as to how far one has to look into future generations to see if there are any future prime factors of the number A (factorization using subtraction). This gives us a way of determining whether numbers are prime or not WITHOUT resorting to the laborious brute force method of traditional factorization.

Since B is a smaller number than A, it can be broken down itself by the same method, therefore only addition and subtraction are needed to determine if numbers have certain factors. This is just one reason why the "magic numbers" are so important, since they have as factors one and only one of ALL of all the prime numbers up to the "nth" prime number. Of course, if there is a prime factor not contained in the listing of prime numbers, then A still may be "factorable". It may have a prime factor larger than any in the list.

This in fact is the basis for "solidifying" the prime numbers in the process of applying the "Generator Function". That is why the prime candidates to the left or negative side of the magic numbers can become stabilized and numbers to the right or positive side of the magic numbers must wait their turn until they are regenerated to the left of a future magic number. That is, the prime candidates to the left or negative side of the magic numbers, up to a point, could not have factors larger than a certain value and therefore can no longer be eligible to have factors "discovered" other than those already tested. They therefore can be guaranteed to be prime. Additionally, these "true" primes are generated in groups (not just one at a time).

Since the "umbrella group" associated with a given "sequential prime product" is a list of "prime candidates" of the given "sequential prime product", it is the list of all numbers that DO NOT have factors in common with the "sequential prime product", or in mathematical terms, it is a list of all "relatively prime" numbers.

On first thought, it may seem cumbersome to keep track of large lists of numbers, some of which may not be prime. However, these give the understanding to the wavelike nature of the primes and are needed to generate the symmetry of the primes around the magic numbers. They also form the lists that are reduced in numbers every time we apply the "Generator Function", and thus are an essential part of the true understanding of the prime numbers in that we now see why some numbers are eliminated while others stay in the final list. And last but not least, they are necessary in the prime number generation process to produce some of the "true" primes. Without them in the "umbrella group", some primes would not be found. There are some incredible features of the "false" primes in the "umbrella groups" including, for example, that every perfect square of every prime number is contained in an umbrella group, and by definition they are all odd numbers (except for $2^2 = 4$).

ADDITIONALY, every "umbrella group" can be written as a simple mathematical expression that guarantees that every prime number is a member of that expression. After just 10 iterations of the Generator Function, we have already identified almost all of the prime numbers that will exist in the future. By using the method of directly calculating the prime numbers, the long laborious super computer calculations will be replaced with a few short arithmetic calculations of addition or subtraction and a comparison to either a table of known "prime candidates" associated with the "sequential prime product" used in the calculation, or the comparison could be made with one of the simple equations that represent the "umbrella group" also associated with the "sequential prime product".

Thus "factorization" takes on an entirely new meaning and the concept presents one of the many pieces of the puzzle needed to solve the riddle of the prime numbers. There are undoubtedly many relationships that will surface regarding all of these numbers, giving future generations of mathematicians a good deal of work to perform. The prime number solution is a beginning, not an end.

X. SNOWFLAKES – SYMMETRY AND 2 DIMENSIONAL PRIMES

In the beginning of the life of a snowflake, it is well known that there is an initial "seed" that begins the crystalline growth that ultimately ends up as the amazingly complex multi-faceted design, no two of which are ever alike. This is the claim at least theoretically. I have never personally tested that hypothesis, although I have shoveled tons of the white stuff.

There is a good deal of 2 dimensional six-sided symmetry in snow flake patterns, and therefore I have chosen these wonderful little pieces of nature's art to demonstrate some of the seemingly random but perfectly formed results of naturally occurring systems, of which the prime number system may play a part.

Notice that 6 is one of the primary "magic numbers" related to the prime number system. We have been dealing so far with prime numbers in the one-dimensional world of the number line and using just the mathematical operations of simple addition and subtraction.

Now looking in two dimensions, imagine a snowflake "seed" in the form of a small dust particle or even a single ice crystal of microscopic size. The seed "grows" and regenerates in possibly the following patterns. In the two-dimensional progression, magic numbers of cells are added each time to give the resulting larger snowflake.

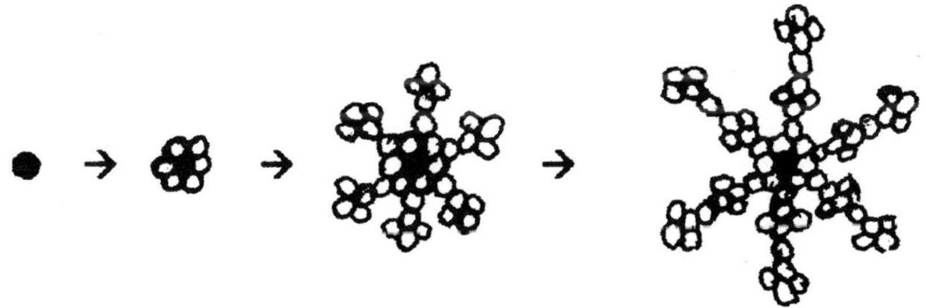

Each successive snowflake has a prime number of circles, and an even number of six sided symmetrical facets.

The final snowflake with 97 cells is shown below.

→ 

The first snowflake seed has one "cell". Add 6 cells and for a total of 7, followed by 37 by adding 30 total cells (another magic number) with 5 cells growing on each arm of the snowflake, followed by 67 (again by adding 30), followed finally by 97 by adding another 30 cells. Could you add 210 more cells and create your own design (210 of course being one of our magic numbers)? This would add 35 cells to each arm. The snowflake would then have 307 cells, another prime number. The point is I generated my one snowflake using a particular sequence of adding symmetrically a "magic number" of cells to always come to a six-sided symmetrical snowflake. Every magic number has 6 as a factor, so adding any magic number of circles evenly and symmetrically distributed amongst the 6 growing arms will produce a unique snowflake.

Now imagine letting 20 people, or even 200 people in a room follow the same progression of adding magic numbers of circles in patterns that they see as "their" snowflake. What would be the probability of finding any two snowflake patterns the same? ... and we have only gone through a very small number of iterations of adding cells to the forming snowflake. This would be a wonderful school project for children to combine art talents with science and mathematics, having them each creating their own snowflake. Using colored pens helps to keep the successive iterations separate and distinct from prior generation patterns. Keeping the spacing equal can be a serious challenge as you can see.

There are of course non-prime numbers that have 6 as a factor, but the point here is that "generation" of naturally occurring systems in nature can take on the same process that we used to generate the prime numbers, and sometimes the prime numbers will be visible in the patterns found in nature. We just have to find them.

In a real snowflake, there are billions of iterations, each one growing and forming on the prior iteration, just as we formed the prime numbers themselves with the "Generator Function". Only this time we are using the building blocks of water molecules in a snowy blizzard high in the clouds above instead of the numbers 0 and 1. Now imagine all the little snowflakes forming in a raging blizzard with varying conditions of atmospheric pressure, temperature and other factors related to snowflake development. Scientists have duplicated the formation of snowflakes in a laboratory to understand the physics of these simple but ultimately complex phenomena of nature. One result is that electrically charged pinheads cause snowflakes to form around the pinheads.

The snowflake is like the galactic structures in that there has to be a formative field emanating from the center of the formation. There can be no other explanation simply based on the observation of symmetry. Although physicists have reproduced the development of snowflakes using electric fields, they have not imagined where these electric fields come from in nature in the extended atmosphere above storm clouds. Further discussion of atmospheric electricity is beyond the scope of this book. However, once again, one can conclude simply based on the properties of repeatability and observed symmetry that the snowflake development process has a controlling generation property that originates in the very center of the crystal.

There certainly must be many formative processes in nature that depend of the growth of systems that could be described using a uniquely defined generator function and possibly even including facets of the prime numbers in the mix. At this point there is a simple but important topic to cover; a future chapter will discuss the fundamental aspect called "repeatability", that is for example, the process and science of forming a snow flake on the far side of the universe is the same as it is here on earth, just as there has to be a repeatable process in the formation of galaxies. A future chapter will discuss other complex systems based on Quantum Mechanics and Genetic building blocks of biological life forms.

XI. THE RIEMANN HYPOTHESIS

Throughout the history of the quest for understanding the prime numbers, numerous analytical equations were developed to estimate what is know as "the density of primes". This means literally, how many primes are there in a certain region of the number line? The ancient Greeks had already determined that there was an infinite number of these unruly numbers. It is one of the first known uses of a mathematical technique called "proof by contradiction". You assume a result and then find a counter example thusly proving your conjecture false.

Many times even amongst professional mathematicians I see this ancient proof misquoted. I will explain and then proceed with more on The Riemann Hypothesis. The Greeks, like modern men, tried to calculate the prime numbers by brute force factorization. But since they wanted to know where to find more, and if there were in fact any more, they made the basic assumption that there were NOT an infinite number of primes. If you multiply all of the prime numbers together that you know about, and add 1, then this number would have to be a prime number ... right? Actually the answer to this is NO, and this is how many people incorrectly repeat the proof used by the ancient Greeks.

The actual proof goes as follows. First assume there is a finite number of prime numbers and that you have found them all. Multiply all known primes together, then add 1 and we will call this number "A". Either the new number "A" is prime OR you are missing a prime number in the list you thought was a complete list of prime numbers, and that new prime number is a factor of "A". By either standard, you have found a new prime number and your original "complete" list really was not complete. You can repeat this process forever, proving that there are an infinite number of primes. This result is paramount to the quest for an equation representing the "density of primes". Without a proof for the uncountable or infinite number of primes, there could be no quest for a density equation.

A density equation cannot predict the exact locations of the prime numbers, they only tell you on average how often you will come across one in your search, that is if you have no other way to determine their values or locations on the number line. If you have not found for a long time, then you are getting closer to finding one by a brute force factorization method.

The quest for more exact density equations lead to more complicated equations and ultimately Bernhad Reimann converted an equation that was used previously and developed a formula related to prime numbers. In levels of abstraction, one would say that it is at least twice removed from the direct calculation of the density of prime numbers. There are many good books and treatises on the exact formulation of this work, but the point is, this initiated a level of complex mathematics that has never been solved. At the very root of this inability to solve The Riemann Hypothesis is the history of not being able to directly calculate the prime numbers. Mathematicians await the next random arrival of super-computer searches and then plug that number in only to be frustrated until the next discovery.

Mathematics requires complete proofs with no possibility of even one exception to the "rule". So the mathematics world waits and hopes that someday the secret of the "density" of primes will be solved based on Riemann's equations. But as noted earlier in this book, what a strange situation exists since now one can directly calculate the prime numbers and determine exactly the density of primes in any interval that you chose. If this had discovered before Reimann's time, then we possibly would not have the Riemann hypothesis at all. But the Riemann Hypothesis and equations have become the heart of many other problems that do have solutions ... but only if the Riemann Hypothesis is correct. The work presented here may hopefully bring the solution of this problem closer to reality. One thing you must understand about mathematicians, they are not hasty to judge results of this magnitude.

There is a $1 Million prize for anyone or group who will provide a solution to Riemann's Hypothesis, but collecting this money is not as easy as one would hope. Even if a solution were discovered today, it would have to wait a minimum of two years before "The Millenium Prize" could be claimed, and chances are it would take much longer.

XII. QUANTUM MECHANICS AND GENETICS

In a random universe with random activity with random rules, could one ever expect the order and amazing conditions we find in the universe around us, from the smallest atomic particles to the largest galactic structures? One night as I was a guest on a radio show with the host who held that there was no divine creator, we slipped into a dialogue regarding this issue. This is something that I have thought about from many points of view including that of a pure scientist and that of a purely faith or intrinsic intuitive feeling of the universe.

I asked the host if there were a water tower in the town where he lived and he said there was. I asked if it were structurally sound and if it might fall. He said it had been there a long time and he did not anticipate if falling, and trusted that it was of sound design and materials. I asked him if someone had designed the water tower and he said there in fact had to be a designer with reputable abilities … of this he was sure. I then asked if he knew the architect or met the structural engineer of knew anyone who ever was involved in building the water tower or even knew from where the materials had come. Once again he did not know. Then I asked the point of the questions … "How can you look at the water tower and be so sure there was a designer and plan and not look around you in this universe and not see the far grander plan that is in progress every day, and of which we know so little and are just beginning to understand the most rudimentary of its secrets?

I often marvel at astronomers who ask if there could be intelligent life outside of own earth or even whether there could be other planets outside of our own star system. Many scientists state that we must wait for discoveries that confirm these, and in the strictest sense of science, this may be correct. Another chapter on "Repeatability" carries this concept further. But for the present chapter I will use the example of the diamond. If you discovered a diamond in the ground, would you wonder if this could be the only diamond in existence? Or would you begin a search to find more? Would you assume that the earth was the only planet in the universe that produced this one diamond? Or would you have to assume that if your found a diamond on the far side of the universe, that it would possess the same properties as the diamond in your hand, and that whatever the process that created the diamond might be, that the process too would work the same way on the far side of the universe? I find this reasoning so fundamental that it remains difficult for me to understand the doubt that some have in imagining intelligent life or other earth like planets. Not only must there be at least one somewhere, there must be so many that you could not begin to count them all.

This brings me to the heart of this discussion and the relationship to Genetics. In the generation of the prime numbers, one can relate the entire understanding of the prime numbers themselves to patterns completely devoid of "numbers", say in patterns of rectangles of objects or concentric circles based on increasing "magic number" for their circumferences. Likewise, one could redefine the entire proof given in this book for the Generator Function in terms of geometric interpretations. Additionally, these representations would be the same on the far side of the universe as they are here. The prime numbers have all of the same properties over there as they do here.

The atoms that make up the nucleotides of basic genetic structures are found in the same patterns, and consist of approximately 30 atoms of hydrogen, carbon, nitrogen, oxygen held to other such units by sugar-phosphate "glue". Do you think these may combine in the same building block forms on the far side of the universe? Especially when they do just that here on earth billions of times over in just your body alone.

As I research more and more with Generator Functions and attempt to apply them to other theoretical structures, there are some striking patterns that emerge. The first is the relative simplicity of the Generator Function and the powerful result of building for example the entire prime number system beginning with just the numbers 0 and 1. Could the genetic code be similar and based on a Generator Function that is so elemental that we will find other biped creatures literally everywhere in the universe, and that this is the norm rather than the exception. The arguments regarding "evolution" and "creation" are actually involved in the same argument when viewed from this perspective. When viewed in this perspective, they are not mutually exclusive concepts.

Extending the analogy of the diamond, no two diamonds are exactly the same just as no two snowflakes are exactly the same, just as no two genetic results would be exactly the same, ALTHOUGH THE PROCESS THAT FORMS THEM ... THE GENERATOR FUNCTIONS ... ARE UNIVERSALLY THE SAME THROUGHOUT THE UNIVERSE.

The one aspect of the Generator Function that is most startling is that "false" primes have to be generated which in turn generate final or "true" primes and then the false starts, so to speak, are eliminated never to resurface again. The importance of these concepts shows why many scientists find dead ends so to speak in their research, and too many stop and look in another direction, when in fact they are possibly just one step from moving past what appears to be the false start.

One personal area of research that I have been involved in throughout my career since my college days, besides the effort to understand the prime numbers, has been the quest to analytically resolve the N simultaneous differential equations related to N-Body systems, which includes everything from Celestial Mechanics and the motion of the stars, planets, moons, etc., to thermodynamics and motions in a solid crystals. It was actually the work in that field that brought me to use the Generator Function and apply it to the prime number problem. The prime number problem is a linear system and easier to manage, but with this solution at hand and new understanding that it brings, there is now hope to apply it to other physical systems.

Regarding Quantum Mechanics, the study of the wave nature of the subatomic world, there is a famous story regarding a chance1972 meeting between physicist Freeman Dyson and Theory of Numbers mathematician Hugh Montgomery. They were conversing research at a meeting. Dyson was studying the nuclear energy levels of heavy atoms while Montgomery was studying the "zero" solutions to the famous Riemann Hypthesis. They noticed that the scaling of the two matched and that there may be a correlation between the prime numbers and the physical system of the atomic nucleus. Ever since there has been a growing rich collaboration, hoping that the physical world may shed light on the mathematical world, and visa versa. Once again, the problem hinges on the problem that the prime numbers have been considered to be "random". The future may change soon now that prime numbers can be understood from basic principles.

XIII. REPEATABILITY IN NATURE

If we were to generally assume that on the far side of the universe, the laws of physics and mathematics are the same as they are here, then there can be only one conclusion. The conclusion would necessarily be that we would find just about the same things we have here, only with that immense variation that comes with differences of time and space, as with species separated by time and space on earth.

We look at past extinction events on earth only to see the rich and varied changes that resurfaced after whatever cataclysmic changes ripped through our seemingly serene little blue planet on the far edges of this smaller than average galaxy in the middle of nowhere.

Change and variation are the norms, but beneath it all lays an underlying plan of simple construction and formal elegance. The basic building blocks are the same, and the rules of engagement are the same. The end results are what differ just as with the snowflake.

The essential property of the prime number system is the repeatability of patterns. These take the form of waves that are symmetrical. Is there a similar property of nature that governs the symmetry in the large-scale structures of galaxies, in other words, how does the one arm of a galaxy know what is going on in the other arm. With too many theories in astrophysics, just looking at the aspect of repeatability would show that certain theories could not possibly be true. How do the arms "repeatedly" end up with such symmetry?

I have asked students in my classes to look at dozens of photos of galaxies. No two are alike. There are warped galaxies, galaxies with unusual bar structures and some with no stars in vast regions with other regions highly populated with dense star fields. Yet, in spite of this vast diversity, if you take a ruler and mark exactly across the center of ANY of these galaxies, you will find the star densities, the warpage, the lack of stars or particularly bright regions and clusters of stars, the one property you always find is amazing symmetry around a central dense nuclear region. The symmetry of galaxies is consistent and is a fundamental property of all galaxies, yet most of the theories that one finds today that pertain to the structure and movement of the arms, cannot explain this simple element of repeatable symmetry.

To my way of thinking, repeatability is as central an idea to mathematics and physics as the laws of conservation of energy and momentum. Yet it is rarely observed or even mentioned. Would one expect to find galaxies on the far side of the universe to develop without symmetry? Then why do scientists develop theories that do not explain it as an essential result of the theory, and more so, if the theory does not consistently explain not only symmetry but also the formation of the dense nucleus, then it has not passed the first and most rudimentary test. So forget about light curves and other data, the theory has not gotten past the first grade. One could in fact possibly simulate one situation in which a given theory could produce symmetry in a galaxy, however, the requirement is to have it work in ALL cases and it therefore must be an essential part of the basic theory ... it must be universally repeatable.

This may seem a bit off point in a book about prime numbers, yet it was this concept of repeatability that so many researchers looked at with the prime numbers, but ultimately came to the conclusion that the prime numbers were random and without an overall structure. On the contrary, when I looked at the prime numbers I assumed that nature in fact had included these essential and subtle properties, but one had to look beyond the first levels to find it. Repeatability is such a fundamental property of all systems that I had to make the assumption that it did exist for the primes ... simple – elegant – subtle – and certainly not obvious, since so many great mathematicians had poured over this problem and did not see the entire solution, although most certainly they all saw bits and pieces.

The point is that repeatability must be an essential part of so many other physical systems. It is really an overlooked commodity since there is really no way to measure or define it in the standard terms that we normally use to define physical systems.

The attached DVD contains the lecture of my January 11, 2007 radio show, which discusses as the main science topic "repeatability". Listen to that show as a supplement to this book. Mathematicians revel in seeing properties such as symmetry, reciprocity, closure, etc. The simple fact that we can discover patterns that are repeatable in the universe indicates that there is a consistency to the basic plan and therefore design. A universe with random events, random rules and random forces would not give the consistent repeatable results we see. When science bases its rules on inconsistent theories, which do not include repeatability as a fundamental starting point, when mixed with repeatable observations, one can only end in chaotic bizarre "science".

XIV. THE WAVE NATURE OF PRIME NUMBERS – HARMONY RULES

Harmonic waves are found in all branches of mathematics and physics. To find them as a basis of the prime numbers is odd since many mathematicians have attempted to study the prime numbers in terms of waves and harmonics and once again, have come to the conclusion that the prime numbers have no patterns and are random at best. Some have seen intermittent patterns but then stated that these end and fail, with no long-term patterns at all.

This book shows that not only are there patterns based on waves, but it shows that there are infinitely many such patterns with an infinite number of wavelengths. Groups of prime numbers are in fact used to generate more prime numbers, which in turn generate larger groups of prime numbers. The concept of past, present and future generations is inherent in the process of directly calculating the complete set of prime numbers. The key to understanding is that the elegance is more than skin deep.

To facilitate understanding of the concept of waves and wavelengths, we will now resort to some basic pattern building that I used long ago just to get a feeling for how the prime numbers behave. Then we will apply the rules of addition and subtraction based on the "magic numbers" to see some very interesting results

To accomplish this take 6 pieces of graph paper and tape the sheets end to end to make a very long continuous sheet (¼" graph paper works well for this exercise). Lay the long paper horizontally on a desk or table. A few rows down from the top, write the numbers starting with 0 in the upper left hand corner and put one number (0, 1, 2, 3, 4, 5, ...) in each successive ¼" box on your graph paper continuing along one row of ¼" boxes until you reach the right hand side of the 6 pages. With 6 sheets taped together you should have numbers listed up to about 240 or more.

Next place an "x" in the ¼" boxes directly under the numbers 0, 2, 4, 6, 8 etc continuing all the way to the end of your long piece of graph paper. Starting in the next row under your first row of x's, place an "x" in the boxes under the numbers 0, 3, 6, 9, etc continuing all the way to the end of the 6 sheets of graph paper. In the next row below this mark an "x" in the boxes under the numbers 0, 5, 10, 15, 20, etc continuing all the way to the end of the paper. Complete one more row by placing an "x" in the boxes directly below the numbers 0, 7, 14, 21, 28, etc continuing all the way to the end of your sheets. You are creating the "repetition groups" talked of earlier in this book, which were used in conjunction with the Generator Function to create the "umbrella groups" containing the "prime candidates" that were then added and subtracted from the magic numbers.

An examination of the repetition group itself does not give a method for calculating prime numbers nor does it give and understanding of the symmetry of the primes. However it is instructive to view the patterns to understand the nature of the "wavelengths" of the "repetition patterns". One could also equate this to the primitive "Sieve of Eratosthenes" used by the Greeks as a basic form of determining small prime numbers, although they never understood the wavelike nature of the primes.

Look at the graph paper and realize that some x's of one pattern fall under the x's of another pattern. For example the row of x's for the repetition group of {2} (your top row of x's) will have an x in the same column as the repetition group for {3} (your second row of x"s) under the numbers, 6, 12, 18, etc. In fact you will see that this pattern repeats every 6 boxes. We say that the repetition pattern of {2,3} then repeats with "wavelength" = 6.

Look for the column where all of the first three rows all have an x under a given number. That number is 30. Therefore the combination of the repetition patterns of {2} and {3} and {5}, when combined to create the repetition pattern {2,3,5}, has a wavelength = 30. This pattern then repeats at 30, 60, 90, etc. When we then use this repetition pattern to build the table of "prime candidates" based on the magic number 30, this repetition wavelength carries over to that part of the process.

Look for the first occurrence where the combined patterns of {2}, {3}, {5} and {7} all have an x under the same number. It is 210, our next "magic number". So the repetition group {2,3,5,7} has wavelength = 210. Now things will get a bit more interesting.

Locate the numbers in the row or {7} in which this row has the only x under a given number. These numbers will be 7, 49, 77, 91, 119, 133, 161, and 203. If you continue beyond 210 you see the same pattern repeating, only added to 210 (e.g. 217, 259, 287, etc.) and if you continued to 420 (which is 2 x 210) you will again see the same pattern repeated only this time added to 420 (e.g. 427, 469, 497, etc). Now take the difference between 210 and the list of numbers 7, 49, etc. The following table lists those numbers.

210 – 7 = 203
210 – 49 = 161
210 – 77 = 133
210 – 91 = 119
210 – 119 = 91
210 – 133 = 77
210 – 161 = 49
210 – 203 = 7

Now go up the middle column and compare to the numbers going down the right column. This indicates a "reciprocity and symmetry". One list generates the other and visa versa. Now divide the numbers by 7. You get the following list of numbers … 7, 11, 13, 17, 19, 23, and 29 (the entire list of prime numbers from 7 to the magic number 30). If you continue this process to the right of 210 you will get the numbers 31, 37, 41, 43, 47 and 49 (49 is the square of 7 and is not prime … so the patterns fail at the square of 7). This shows that you have generated all of the prime numbers up to the square of 7 (7 is the largest prime number in the magic number 210). Not only did you directly calculate the primes, but they are symmetrical and the inverse of the process generates the same group of numbers but in reverse order. Please ask anyone who says there are no patterns in the prime numbers or that the prime numbers are random, to please continue this process with every "magic number" until they are convinced otherwise.

The patterns we just saw are inherently blended into the smooth operation of the Generator Function which directly calculates the prime numbers in groups. The properties of the primes include symmetry, reciprocity (one group of primes generates another group and the second group in turn can generate the first group) as well as "closure" (the property that all primes are calculated, with none left behind such that no "outsiders" are included in the final mix).

Thus the prime numbers form a unique very interesting mathematical structure. In the graph paper that you have been putting x's, if you continue with the next prime number 11, your "wavelength" of the repetition group {2.3.5.7.11} = 2310. When we first started using examples in the beginning of this book you started looking at the prime number table using the magic numbers 30, 210 and 2310. Look at the prime number table now and once again look at the patterns of prime numbers as they fall to the right and left of these numbers. Calculate the differences between the primes and the magic numbers. Then once again go to multiples of these such as 60, 90, or 270, which are multiples of 30. Then go to 210, which is also a multiple of 30 but has an additional number 7 in its set of prime numbers making up the "sequential prime product". Look at what happens when you add 7 to each of these numbers. 37, 67, 97, and 277 are all prime numbers, but 210 + 7 is not prime since 7 is in the sequential prime product of 210 (e.g. 210 = 2 x 3 x 5 x 7), likewise any multiple of 210 added to 7 will NEVER be prime. Written in mathematical terms that would be any number of the form

$210\,n \pm 7$ will never be a prime number

Understanding the prime numbers and why they are in the table as well as why other numbers are not in the table is the result of calculating the prime numbers directly. If the only result were the ability to calculate prime numbers directly, that would be a major achievement, however, the understanding of each and every number and its reason for being in the prime number table or not as well as understanding the many patterns is an equally significant result. The concepts of wave patterns are central to this understanding.

XV. OLD METHOD OF DETERMINING PRIMES

If you have a great deal of time available, attempt taking any random large number of 50 digits or more and using a calculator, attempt to find prime factors using the brute force method of repeated division. You do not have to repeat the exercise, but this is just to give the idea of the extensive amount of number crunching needed to determine just one prime number, or if your number is not prime, all of your calculations are for a null result and you get to start over with the next larger number. In addition to endless amounts of extensive labor, no real understanding is gained other than the satisfaction of possibly finding a new prime or prime pair. But that elation is short lived as you return to the calculator to search for the next prime that is larger and more remote than the one you just discovered.

The computational efforts to arrive at prime numbers using the new direct calculation method will be very different from the traditional method of factorization. The amount of computation required based on the traditional factorization method has diminishing returns on every aspect of the process. As the potential prime numbers get larger, the process is complicated by the fact that almost every number has to be tested, the number of factors increases as the numbers get larger, the number of primes needed to be included in the test grows, and the density of primes goes down, so there are increasingly more negative result tests. It is a process of diminishing returns.

The new method of locating primes shows that the prime lists are stabilized relatively quickly, the candidates are never "tested" by dividing successive legions of prime factors, but are directly calculated from an ever decreasing list of "prime candidates". Additionally, patterns and relationships can be programmed into the search for identification of designated prime associations.

The elimination process works very quickly and you will see that before too many iterations of the application of the Generator Function, almost all of the prime numbers all the way out to infinity will already have been identified with only a few percent remaining to be eliminated. This will demonstrate the power of this method relative to the standard factorization.

To give an idea of the rapid isolation of the majority of prime numbers using the Generator Function method, it will be instructive to calculate the percentages of prime numbers eliminated by using the repetition functions discussed in the previous chapter. The repetition pattern {2} eliminates 50% of all numbers. One would possibly think that combining the repetition group {2} with {3} would take away one third of the remaining numbers, but a closer look shows that, since every other number in the {3} repetition group coincides with the members of the {2} group, then the {3} group only eliminates on out of every 6 numbers on the number line from being prime. Combining the {5} repetition group only eliminates 2 numbers out of every 30 (see the x's on your pages of graph paper to count this and verify this result). The {7} group only eliminates 8 numbers out of the 210 wavelength, etc. So the calculation for the number of non-prime numbers eliminated by this rudimentary method is

1 - 0.5 – 0.1666 - 0.06666 - 0.038 - etc.

Each term gets increasingly much smaller as the wavelength of the repetition group gets larger.

These examples illustrate some of the properties of prime numbers and their density as one moves out towards infinity along the number line.

XVI. THE FUTURE OF PRIMES

In the past and prior to the solution of directly calculating the prime numbers, the worldwide mathematical progress focused on extremely complex mathematical proofs and formulas that only a few people in the world could really understand. As with many branches where people with vast levels of training work in hallowed halls or corporate settings, it can become a veritable Ivory Tower, with terms, international conferences and published papers that just a few people read and fewer people would fully comprehend.

Prime numbers have been possibly one of the most esoteric of topics. As already noted, over 100 years ago mathematicians gave up attempts to directly understand the prime numbers and began down a path so complicated that fewer than few professionals truly had a grasp on what the current state of the art might be. There are teams of mathematicians and physicists that regularly meet to discuss for example just "The Riemann Hypothesis" and the new developments that may lead to its solution.

The dichotomy of one new discovery, such as the one outlined in this book, changing the rules as they stood over 100 years ago, is somewhat strange to say the least. One recent set of papers even went to far as to say that a proof had been completed stating that the prime numbers were random. This certainly put the final nail in the coffin for anyone to initiate a search for a direct calculation solution. No sooner was the ink dry on that paper than in fact the present work was solidifying and being prepared for public distribution. The present work shows the opposite, that in fact the prime numbers are anything but random, and with the proper training even grade school children can learn to manage and understand why every prime number is in the table, and how it relates to an infinity of other prime numbers. It also shows amazing patterns and mathematical properties such as symmetry. Before the present work, who would have ever thought that the primes possessed symmetry as a fundamental property?

So there is a true but uncanny quagmire now in the world of the prime numbers. Could some of the other work of the past 100 years also contain flaws? This would certainly raise a nightmare situation in the world of mathematics. Be that as it may, some will continue on the path of "The Riemann Hypothesis" and the complex proofs regarding the density of primes. The solution to this problem may be at hand in fact before this book actually gets into the hands of the public. But remember that incredible amounts of materials exist that require the solution to The Riemann Hypothesis and therefore the search may hasten at this point. One of the most perplexing problems in sub-atomic physics depends on a solution to equations based on the prime numbers. And then there is the search for prime pairs and the infamous "Mersenne Primes", special prime numbers that occur in pairs based on a mathematical relationship.

For others, prime numbers will take on an entirely different direction, filling hours of time playing with the patterns now obvious in the primes, and developing the myriad of associations that even I have not imagined at this point in time. These may relate to real geometrical and physical entities found in nature or the mathematical equations that describe them, from the strange world of subatomic particle physics to genetics and possibly even to the largest structures in the universe.

Prime Numbers are truly the building blocks of the number system, and in fact are an independent set of numbers totally independent of the non-primes. This fact alone will cause untold mathematical progress in the form of "Algebra of the Prime Numbers". This is just one area that undoubtedly will be a major area of study by mathematicians, to discover and build the algebraic structure of the prime numbers. The current book only lays down the initial foundation for that huge area to open.

I can envision all of these, but it will take extensive efforts by people taking this as a specialty to truly unlock the complete set of mysteries of the primes, now that they can be directly calculated.

And last but not least are the young school children who will now have a new chapter in their text books ... "Calculate Primes". Not only will these be studied in the grade school level, but the young students will also gain an understanding of how the "false" primes must be part of the scenario to in turn find the "true" primes and that this may be a fundamental way that nature works.

XVII. ENCRYPTION – IS IT DOOMED?

As promised in the pre-announcement of this book, an entirely different but relatively unknown side of my past is being revealed with the release of this book. A good half of my career has been spent in the world of computer design and telecommunications. For over 10 years I worked as a Principle Systems Engineer designing every aspect of secure computer communications protocols, hardware and software and making the two worlds merge. My physics as well as computer background has been at work in areas of solid-state physics and computer design and performance analysis.

I have never been a programmer; yet I have designed every aspect of micro-code and software from individual functionality and protocols to instruction sets for computer processor chips. I have worked with teams of engineers on both the hardware and software ends of computer communication with everything from microchips to extensive satellite communications systems and large scale private networks for some of the world's largest corporations. I rarely discuss these aspects of my past, as they were not relevant to my more public life dealing with my theoretical science work for which the majority of people know me, and which has sprung such heated controversy and personal attacks on me and my good name.

Likewise, I rarely talk about specifics of much of my professional past regarding places where I have spent large amounts of time, including work with high level Russian scientists and teaching outside of the USA. These I still keep private to allow me to continue to communicate with long term associates in private and without the complications of public scrutiny or that lovely group of government agents that attempt to track and sidetrack my every move.

So regarding the topic of computer communications, I possibly know as much about this topic as most people you might meet. So the topic of encryption is near and dear to my heart. The simple fact is that a good deal of secure telecommunications and other computer protocols depend on some form of encoding. The most secure of these have depended on either random number generation or the use of large prime numbers. I have even designed electrical encryption techniques, some of which I have never marketed and employ for my own personal use.

To my way of thinking, the methods used today in international banking and Internet marketing is fraught with flaws, and I am amazed that more people have not invaded highly secure systems and "taken them to the cleaners" as they say. The entire problem is not one of security, as these systems change regularly and in fact for that reason alone are hard targets to crack. The true problem is that if anyone did crack one of these secure communications facilities, the assumption would continue for an indeterminable amount of time that the system was still secure, and the damage could grow to insurmountable levels, if you see what I mean. The perpetrators could dust off their tracks and no one would be any the wiser until the "crooks" were long gone.

So, by publishing the method to rapidly calculate large prime numbers, one aspect of encryption codes may come one step closure to obsolescence. In my more than professional view, I am doing the entire industry a HUGE favor. They need to develop real encryption systems that are in fact 100 % secure, not 99.99999% secure.

Encryption systems usually work on a system of "keys" in the form of large numbers. Like 2 keys to a safety deposit box, both keys are needed to open the box. Using safety the deposit box as an example, the bank keeps one key in its vault, which is a master key for all of the safety deposit boxes. The customer has his own key that fits his box alone. Both people have to be present to open the box and there is additionally a recognition and sign-in required. All this seems quite secure until you enter the world of computer communications, counterintelligence and wire-tapping. In short, NOTHING you send over the Internet INCLUDING on secure links is really 100% safe.

The typical person however is saved since you are just one of trillions of transactions occurring every second of the day. The problem of course comes in when someone targets you specifically. We hear regularly how little Johny somewhere playing with his home computer was able to hack into the most secure military computing system and play war games or something. We hear how the Chinese regularly tap into major US corporation and military databases and pilfer information. They have entire teams of people working around the world engaged in these activities.

US federal government pressure has required that private companies who develop software, computer equipment, firewalls or encryption software algorithms keep “backdoors” in their software and computer hardware to allow federal agencies an opening to get into computer systems. Companies or groups that do not comply have faced “consequences”, even though there is really no law that oversees this world of “backdoor computing”. These many times are the entry points hackers use to get into computing systems. Entire stories of international intrigue and espionage have surfaced regarding software designed for one purpose and then used or obtained by another group. This topic is somewhat of a diversion from the main topic of this book. However, it is possibly one of the most relevant topics to some people as this book is released to the general public

So ask me about encryption? Yes I know a few things about it, but the one thing I can tell you is that if it is “on-line”, it is available to intrusion by someone somewhere. We are living in a very tenuous world of communications, where smoke and mirrors lurk around every corner, where powerful search programs probe the internet constantly looking for somewhere to enter ... someone’s identity to borrow or someone’s computer to infect ... sometimes just for fun ... sometimes for more nefarious reasons.

So in advancing a mathematical solution that may potentially break down one of the currently used encryption techniques, my attitude is that I am doing you all a giant favor. At least I have the decency to make this information known, and not use it for harming the public.

The primary reason for the patents and trademarks regarding this book and its contents involve the issue of using this material to calculate prime numbers and use them in such areas as encryption. As an invention, the process presented in this book (and numerous extensions of the method) constitutes a patentable item.

Thus the patents and trademarks are the one mechanism to retain control of this process that was kept secret during the time that this book and DVD were under production to distribute this information to the public. See Chapter IV for reference information on legal issues.

The only allowance noted in the Chapter IV on COPYRIGHTS – PATENTS AND TRADE MARKS has been to allow legitimate teachers (note: for K-12 only) to freely copy materials from this book to facilitate classroom presentations for the direct benefit of young students attending schools (note that this does not include “classes” in corporations or other venues that some may consider “classrooms”. It only applies to legitimate K-12 classroom teachers). Note that the

DVD cannot be duplicated since an entire class can watch the DVD without being duplicated, and allowing DVDs to be copied opens entirely different issues.

The primary reason for the patent and trademark additions to what normally would just be a copyrighted book is to control the process of directly calculating primes as well as the growing list of options and additions, with possibly the greatest application of this being in the area of encryption and digital communications. As noted, this book is not the venue to release the patent and trademark information since it will increase and multiply and this book will remain the same printed matter as time passes.

Regarding encryption, there are other ways to build up these secure systems and this is a heralding possibly that the entire encryption industry should see a revival and resurgence of new techniques that are long overdue.

The next chapter is a reprint of the original mathematical treatise dealing with this topic. It was originally passed to certain key people quite some time ago (before this book and DVD were being produced) under strict controls to register the work and allow this public version to be released. It has been in the hands of key people who would recognize and register its time of origination and authorship, and was used for other registration purposes to protect the work.

XVIII. APPENDIX I – THE ORIGINAL MATHEMATICAL TREATISE

Direct Calculation of Prime Numbers

by J. M. McCanney

Abstract

This paper defines a new concept for calculating prime numbers in groups based on certain identifiable key numbers. Subsequent papers, which are not part of this submission, deal with more complex issues such as the density of primes, prime patterns and the generalization of these concepts to topology, etc. In this first paper, using only the numbers 0 and 1, and addition as the primary operator, the entire range of prime numbers can be calculated. New concepts and relationships between the primes are developed. A new Alternative Definition of Prime Numbers is advanced which is equivalent to the standard definition, but allows a deeper understanding of prime numbers. Traditionally, primes are thought to be simply a set of numbers that do not fit into the orderly multiplication table of the non-primes. To the contrary, the primes constitute an elegant set of independently self-consistent numbers with properties such as closure, completeness, symmetry and reciprocity (where groups of primes are used to generate other groups of primes and the second prime group can be used to generate the first group). The primes form a closed group and do not require the standard multiplication process (e.g. factorization) to generate or test them. The complete set of primes is generated using the new definition (with only addition and subtraction as operators) and natural "eliminator" rules. Once these rules are understood from basic principles, the prime numbers can be calculated with confidence and without reference to the non prime numbers and their standard tests of calculating factors. The completeness of the entire set of primes is guaranteed by this algebraic structure. Additionally, any number can be tested to determine if it is prime or not, using the same definition used to generate the complete set of prime numbers, using only addition and subtraction.

- Footnotes: Classification 11 Number Theory
- Keywords: Number Theory, Prime Numbers

Introduction

A number of important basic concepts are first introduced to lay the framework for the "generation rules" and "eliminator rules" which follow naturally from the new Alternative Definition of Prime Numbers. From basic principles, a number of new concepts are developed which require definition of terms such as "repetition patterns" and "repetition groups", "sequential prime products", "ancestor" and "parent" and "child" "generation groups" of primes, "prime candidates", "umbrella groups", "negative and positive directions" for the generation of primes, "symmetrical" and "asymmetrical" primes and the "reciprocity" of primes.

Prime numbers are generated in groups and it is shown that all prime numbers have their ancestry in the numbers 0 and 1, thus all future generations of groups of primes will have similar and repeatable patterns based on previously generated groups of prime numbers. Examples will be given and it will be shown that the repeatable nature of prime patterns is a subset or natural byproduct of the overall concept of prime generation presented here. Not only are the prime numbers generated to infinity, but also a deep understanding of the prime number system is advanced.

This is the first application of a new mathematical operator called a "generator function" which operates on a basic primal set of elementary building blocks (in this case the numbers 0 and 1). The same operator is applied repeatedly to each successive solution, which provides a final result for a complete solution. This method holds promise to solve larger scale problems such as quantum mechanical systems, building molecules or larger constructions such as DNA and solving N-Body problems in Celestial Mechanics. The solution of the prime number problem presented in this paper is the first application of this method.

Definition of Terms and Basic Principles

The following discussion is necessary to develop the importance of "sequential prime products" as the key numbers around which the prime numbers are generated.

Establish on a number line the patterns of all multiples of 2, then all multiples of 3, then all multiples of 5, etc. These "repetition patterns" collectively form "repetition groups" that eliminate numbers on the number line from being considered as prime, using the traditional "factorization" definition of prime numbers. In selecting the next number to generate a "repetition pattern", one need only select already identified prime numbers since, for example, all multiples of 4 have already been eliminated by the "repetition pattern" based on 2. This could be used as a method for determining prime numbers, but using this method would not reveal the true and subtle nature of primes.

What is important from this discussion of "repetition patterns" is that one can define "repetition groups" R_{gn} (where n is the nth sequence "repetition group") based on a sequence of primes and their respective "repetition patterns". The first group $R_{g1} = \{2\}$ is made up of the number 2 and its "repetition pattern". Defining the second and subsequent sequences, $R_{g2} = \{2,3\}$ is made up of 2 and 3 and their combined "repetition patterns", followed by $R_{g3} = \{2,3,5\}$ and next $R_{g4} = \{2,3,5,7\}$, etc. with their combined "repetition patterns", etc. Using {2,3}, for example, this "repetition group" has a "repetition group length" of 6 and repeats the same pattern indefinitely. The "repetition group" $R_{g3} = \{2,3,5\}$ has a "repetition group length" of 30 and again repeats indefinitely, with the {2,3} "repetition group" of length 6 being a subset of this larger "repetition group". Another way of looking at this is that the "repetition group" {2,3,5}, for example, converges on the numbers 30, 60, 90, etc. The "repetition group lengths" are 2, 6 (e.g. 2 x 3), 30 (e.g. 2 x 3 x 5), 210 (e.g. 2 x 3 x 5 x 7), 2310 (e.g. 2 x 3 x 5 x 7 x 11), etc. The first 9 numbers are listed in Table 1.

nth sequence	largest prime number $p_{max\,n}$	"repetition group length"
1	2	2
2	3	6
3	5	30
4	7	210
5	11	2,310
6	13	30,030
7	17	510,510
8	19	9,699,690
9	23	223,092,870

Table 1

(Each "repetition group length" represents a unique set of primes based on the periodicity of its "repetition group"; the length is always an even number since 2 is a member of every group; note that after 6 they all will end in 0 since 5 and 2 are members of all of them. The concept of "repetition group length" will become important in advanced topics. Note also that every "repetition group" is a subset of the next or future generation "repetition group".)

This gives rise to the fundamental concept of "sequential prime products". These are equal to the "repetition group lengths" of the "repetition groups" of primes beginning with {2} up to and including the nth prime {2,3,5, ... p_n}. The term "sequential prime product" will be used from this point forward and is represented by the following form:

$$S_{pp\,n} = \Pi_1^n p_i = \text{"repetition group length"}$$

(Where p_i is the ith prime number of the nth "sequential prime product" $S_{pp\,n}$).

For example ... $S_{pp\,5} = \Pi_1^5 p_i = 2 \times 3 \times 5 \times 7 \times 11 = 2{,}310$

It will be seen that these in fact are the numbers, which are pivotal to the prime number system. It will be shown that the primes can be generated in both the positive and negative directions from the "sequential prime product" numbers and are the center points for a new concept termed "umbrella groups". An "umbrella group" U_{gn} is the nth "umbrella group" associated with the nth "repetition group" R_{gn} and its associated "sequential prime product $S_{pp\,n}$. Members of U_{gn} will be called "prime candidates", with p_{cj} being the jth "prime candidate" in the "umbrella group" U_{gn}. The final set of true prime numbers is a subset of this group. The operator will eliminate the "false primes" and discover all of the true primes in groups, with each application of the operator defined below, using only the mathematical operators of addition and subtraction. In brief, it provides the long sought formula for directly calculating the prime numbers.

The definition of "umbrella group" is given on the next page. This constitutes the basis for the new Alternative Definition of Prime Numbers. By definition, the members of U_{gn} are symmetrical around the value of the associated "sequential prime product" $S_{pp\,n}$ because the members of the previous "repetition group" are added to and subtracted from the "sequential prime product" associated with each umbrella group. Starting with just 0 and 1 and the operation of addition, the symmetrical "umbrella groups" are generated, each "umbrella group" U_{gn} being generated from the previous "umbrella group" $U_{g(n-1)}$. Inherent in the definition is the "eliminator rule" spoken of earlier, which eliminates the use of "prime candidates" p_{cm} of the previous "umbrella group" $U_{g(n-1)}$ that are members of the "repetition group" R_{gn} associated with the nth "sequential prime product" $S_{pp\,n}$. Each successive "umbrella group" will be a subset of the previous "umbrella group". These act as "waves" of periodicity $S_{pp\,n}$, modifying the previous "umbrella groups" and reducing the number of "prime candidates" in each successive "umbrella group". The use of these terms and their meaning will become clear when the examples are used to directly calculate the prime numbers.

Eventually some "prime candidates" are eliminated by these natural "elimination rules", while the others remain and are "stabilized" as true prime numbers. The rules, which guarantee that a group of primes has been "stabilized", are a natural by product of this method of generating prime numbers. Additionally, the proof that the generated and stabilized groups of primes are complete is also a natural by product of this algebraic system. The prime numbers are generated in groups. This is why primes continue to show the same patterns, as they are the direct result of the patterns already established in "ancestor groups" (previously generated groups of primes) that go back to the original numbers 0 and 1. This will be explained in more detail when actually developing the prime numbers below, as well as why these patterns vary as the primes are generated out to infinity. The implications of the new Alternative Definition of Prime Numbers will become more apparent in the solutions of other related problems of mathematics. The new definition follows from the most basic principles of Number Theory.

The notational representation for an "umbrella group" is as follows (the symbols ε and $\underline{\varepsilon}$ represent "are an element of" and "are not an element of" respectively):

$$U_{gn} = \Pi_1^{\,n} p_i \pm 1 \text{ and } \Pi_1^{\,n} p_i \pm p_{cm} \,,\; p_{cm} \,\varepsilon\, U_{g(n-1)} \text{ and } p_{cm} \,\underline{\varepsilon}\, R_{gn} \text{ (for all } p_{cm} > 0)$$

(The special use of $\pm$ 1 will be explained below.)

The fundamental concept, which will be the basis of "generation rules" and "eliminator rules", is that the "sequential prime products" and their related "repetition groups" are essentially eliminating numbers from being primes (these would be the non-prime numbers found in standard multiplication tables). The only remaining "prime candidates" that could possibly be prime numbers are all numbers that remain to the positive and negative directions of the selected "sequential prime product" $S_{pp\,n}$ that are not members of the "repetition group" R_n associated with the nth "sequential prime product". This is the "umbrella group" U_{gn} with its members the "prime candidates" p_{cj} that are symmetrical around the value of $S_{pp\,n}$ (for all values > 0). Not all of these "prime candidates" will ultimately be primes.

For example, given the "sequential prime product" $S_{g3} = 30$, based on the "repetition group" $R_{g3} = \{2,3,5\}$, adding or subtracting any member of the "repetition group" R_{g3} to 30 will produce a non-prime number. Thus the only numbers that could be "prime candidates" are found by adding and subtracting to 30, the numbers 1 and all "prime candidates" discovered up to this point by the prior "umbrella group".

$$30 \pm 1, 7, 11, 13, \ldots \text{ (for all values > 0)}$$

Note that 2, 3 and 5 are not in the group added to 30 since they are members of the group $R_{g3} = \{2,3,5\}$. The Alternative Definition of Prime Numbers will include this topic. This generates the symmetrical "umbrella group" of "prime candidates" around the "sequential prime product" $S_{pp\,3} = 30$. The "eliminator rules" will then be used to eliminate some of these "prime candidates" as future "umbrella groups" are developed, but one never tests these remaining numbers for factors as is usually done to determine if a number is prime or not. The tests depend only on addition and subtraction as operators. This "umbrella group" will have periodicity 30, that is, the patterns of "prime candidates" will repeat with a "group repetition length" or "wave length" = 30.

The new test used to determine if a number is prime or not is performed by allowing the waves of successive "umbrella groups" to modify the various members of prior "umbrella groups" of successive "sequential prime products". The future "umbrella groups" modify and correct the prior "umbrella groups". This operator generates true primes up to a guaranteed value, and is then reapplied using these in addition to newly found "prime candidates" to add and subtract from the next higher order "sequential prime product" to generate the next set of true primes. This operator not only generates and identifies all the prime numbers, including special classes such as Mersenne Primes, but also shows the elegant structure of the primes. This closed algebraic group will demonstrate why it is fundamentally important to begin with just 0 and 1 and proceed in the manner described.

In actually developing the primes, "asymmetrical primes" will be discovered in which, for example, by adding and then subtracting the same "prime candidate" to a given "sequential prime product", one of these will discover a true prime and the other will not. These asymmetries reflect the "eliminator rules" that eliminate certain "prime candidates" from originally generated "umbrella groups" to give the final "stabilized" set of primes ... the true primes. This will become apparent when developing the primes below. (Note that in selecting the name for "eliminator rules", the term "selection rules" was also considered and could be maintained by some to be more appropriate.)

The most elegant aspects of this representation of the primes are the principles of symmetry and reciprocity. As one generates the primes from one of the "sequential prime products" in the negative direction, they coincide with the primes that were already generated from an earlier generation "sequential prime product" (with the primes generated in both the positive and negative directions), but with some special exceptions. These exceptions are eliminated by the newer (higher order) "umbrella group" to give the final and complete set of prime numbers, up to a certain determinable value based on the prime numbers that make up the "sequential prime product".

Additionally, no matter which "sequential prime product" is used to begin the prime generation process, and no matter in which direction one proceeds (e.g. positive or negative direction), the primes in general turn out to be the same. That is, the process of generating primes and "umbrella groups" by addition and subtraction of already identified "prime candidates" from prior "sequential prime products" will discover all of the primes. All "umbrella groups" contain all of the prime numbers as a subset. Each successive "umbrella group" will be a subset of the previous "umbrella group", and therefore will also be a subset of all prior "umbrella groups". As already noted however, it is essential to begin with just 0 and 1 as the starting point. Symmetry, reciprocity, completeness, periodicity and closure are essential properties of this method of generating prime numbers.

These uncanny and unexpected results resolve many of the inherent mysteries of prime numbers when viewed in this new structure. In this respect they are a totally independent number system ... a closed group with completeness, symmetry and reciprocity. They are symmetrical around the "sequential prime product" numbers since one finds the primes by adding and subtracting already discovered primes from the "sequential prime product" numbers to form the respective "umbrella groups". The "umbrella groups", and therefore the primes, have guaranteed reciprocity in that all primes are members of all "umbrella groups" generated to the positive or negative directions of all "sequential prime products". That is, all "umbrella groups" will in fact discover all the primes. In viewing this on the number line, the primes generated and discovered in the infinite number of "umbrella groups" all match, even though they have their origins in seemingly totally disjoint numbers (the "sequential prime product" numbers). These properties are unexpected but most assuredly are the most amazing aspect of the prime number system. The prime numbers are well connected in that they are all intimately related in their respective "umbrella groups"; all being formed from the original source of numbers 0 and 1. There are fundamental reasons for these properties.

Alternative Definition of Prime Numbers

The traditional definition of a prime is a number, which only has 1 and itself as factors. An alternative definition will prove to give the same result, but eliminates the need for multiplication and factorization as a basis for identifying primes. The new definition is as follows: a prime number is a number not eliminated by "eliminator rules" in the sequential development of the prime number system, determined by adding and subtracting 1 and already identified "prime candidates" of the previous "umbrella group" $U_{g(n-1)}$ (except for all members of the "repetition group" R_{gn}) to "sequential prime products" $S_{pp\,n}$ (for all values > 0). The traditional and new alternative definitions will give the same results, but are inherently very different. To accommodate the special and initial case of 2, the definition uses the concept that primes are those numbers that initiate "repetition patterns" on the number line. This definition could be universally used, but as already noted, would constitute little more than already exists using the traditional multiplication and factor definition of primes.

The final result is that one will be able to look at the table of prime numbers and identify why each prime number is present and why other numbers are not in the table based on the differences between the number and any one of the infinite set of "sequential prime products". Examples will be given at the end of the paper, but the primary result is that the prime numbers are not a random set of numbers lost from the standard multiplication table. They are in fact a closed group of highly related numbers with elegant properties such as symmetry and repeatable wave structures.

Development of the Prime Numbers from 0 and 1

In developing the primes, the use of multiplication per se is never used. The numbers are generated using just addition (and subtraction) as an operator. The nature of primes comes completely from the selection rules called "eliminator rules" in the process of adding and subtracting 1 and already identified "prime candidates" contained in the previous "umbrella group" to "sequential prime products". Note also that multiplication is not used in determining the values of the "sequential prime products" (even though they are called products), rather they are determined by the "repetition patterns" and resulting "repetition group lengths" of primes that have already been generated by basic addition starting with just the numbers 0 and 1.

This subtle difference between "repetition groups" and multiplication may not be so apparent at first, but will be far more important early in the process of generating the primes and more so in future applications when extensions of these principles are used in multidimensional topology (for example when used to generate spatial solutions). Other differences will become apparent, for example, when a new method of "factorization" of any number is defined by taking the difference between any number and a "sequential prime product", that is, factorization will be redefined in terms of addition and subtraction relative to the base points of the prime number system (the "sequential prime products").

Now to generate the primes ...

Start with the numbers 0 and 1 and addition as an operator. Using addition generate the Natural Numbers as would normally be done in Number Theory. Generating the number 1 + 1 = 2, since this is the first number besides 1, it is prime. This is based on the alternative definition of primes given above. Next, form the "repetition group" based on 2. These numbers would normally be called "multiples of 2" but in this definition they are called the members of the "repetition group" of the first "sequential prime product" which is $S_{pp\,1} = 2$.

The process of adding 1 and all other known prime numbers (a null set) to 2 forms the next set of "prime candidates" which constitute the "umbrella group" of the "sequential prime product" $S_{pp\,1} = 2$.

$$U_{g1} = \Pi_1{}^1\, p_i \pm 1 = 2 \pm 1 = 3, \text{n/a (1 not used)} \;\text{ and }\; \Pi_1{}^1\, p_i \pm p_{cm},$$
$$p_{cm} \varepsilon\, U_{g0} \text{ and } p_{cm} \underline{\varepsilon}\, R_{g1} \;\; (\text{for all } p_{cm} > 0)$$

Since there are no other "prime candidates" identified at this time, 3 is the only new "prime candidate" discovered using the "sequential prime product" $S_{pp\,1} = 2$. By definition, 3 is a member of the "umbrella group" U_{g1} of the "sequential prime product" 2. Since 3 has now been identified as a "prime candidate", it can be added to 2 to generate more "prime candidates". 2 + 3 = 5, so 5 can now be added to 2 to give 7, etc. to infinity. This process continues, but almost immediately it seems to break down, as 2 + 7 = 9, which we know is not a prime. This is where the next sequence will display how "eliminator rules" work. 9 will be eliminated when generating the next group of "prime candidates" in the next "umbrella group" associated with the next "sequential prime product". In future "generations" of "umbrella groups", sometimes this elimination will not occur with the next generation "umbrella group" but will require other future generations to eliminate the "non-prime" from the current "umbrella group". The reasons for this will become obvious as these cases arise. Calculating the limit on how far into future generations this may occur is relatively simple, but the generalized proof of this is the subject of other unreleased papers on this topic, dealing with topics such as the density of primes and other more complex related problems.

Care has been taken to identify the difference between the prime number 2 and the "sequential prime product" $S_{pp\,1} = 2$. This is the only case in which the prime number and the "sequential prime product" could have the same numerical value. Note as the process proceeds that the criteria to be prime are based on the future "umbrella groups" and naturally occurring "eliminator rules". These will modify this "umbrella group" until it ultimately is stabilized for the remainder of the process of discovering all of the prime numbers.

A very important aspect of "prime candidates" generated by this method must not be overlooked. That is, all "umbrella groups" extend to infinity in the positive direction, but only to 0 in the negative direction, with the "sequential prime product" as starting point. So the umbrella group U_{g1} generated by adding all the consecutively generated odd numbers to 2 should not stop as we did above with the "prime candidate" $p_{c4} = 9$, since somewhere between 9 and infinity, there will be infinitely many primes that in fact will be odd numbers, and will be in the "umbrella group" of the "sequential prime product" $S_{pp\,1} = 2$. All prime numbers are a subset of U_{g1} and all future "umbrella groups". This concept will become increasingly important as larger more complex "umbrella groups" are formed around future "sequential prime products". With this said, U_{g1} is defined as:

$$U_{g1} = 3,5,7,9,11,13,15, \ldots$$

Note that 2 is not a member of U_{g1}, since it is a member of the "repetition group" R_{g1}. The group has periodicity = 2 and contains all the prime numbers as a subset. Future "umbrella groups" will modify and correct this group to refine and eventually "stabilize" the true primes from this group. There will eventually be a generalized "stabilization rule" that all "prime candidates" $p_{cj} < (p_{max\,n})^2$ will be true prime numbers, where $p_{max\,n}$ is the largest prime number used to generate the "repetition group" R_{gn}. In the current case of n = 1, $p_{max\,1} = 2$. The generalized proof of this (with variations) will be a topic of one of the associated papers and is a complex topic. But for the purpose of this paper, the members of U_{g1} that are less than $(p_{max\,1})^2 = 4$ can no longer be modified by any future "umbrella group". Applying this criterion to the above list (and including 2 from the prior list of known stabilized primes) 2 and 3 are "stabilized" or true primes and can be used to generate future "repetition groups" and their associated "sequential prime products". In general, as this process proceeds, there will be far more "stabilized" primes available for the guaranteed generation of all future "repetition groups" and their respective "sequential prime products".

Another way of looking at this is that to the left or negative direction from a given "sequential prime product", the list of primes stabilizes and becomes fixed within a determinable and fixed range of future "umbrella groups' working to eliminate non-primes from the list. However, looking in the positive direction, the pattern of any given "umbrella group" will repeat with periodicity S_{gn} and emerge repeatedly (in fact an infinite number of times) to be congruent with future groups of primes. This is why the patterns of primes seem to continually rise and fade like waves as the primes continue out to infinity. The observation of these repeating patterns is a direct result of the process of the generation of primes using the concepts presented in this paper. Remember that all future generations of prime groups and their patterns are reflections of already existing "ancestor" prime patterns, as they overlap repeatedly, and as "umbrella groups" are generated from prior "umbrella groups" around the never-ending succession of "sequential prime products".

Next generate the subsequent "sequential prime product" and its related "umbrella group".

Using 2 and 3 generate the next "sequential prime product" S_{g2} , and follow the generation rules for creating the "umbrella group" U_{g2} of "prime candidates".

$$U_{g2} = \Pi_1^2 p_i \pm 1 = 6 \pm 1 = 7{,}5 \text{ and } \Pi_1^2 p_i \pm p_{cm},$$
$$p_{cm} \varepsilon U_{g1} \text{ and } p_{cm} \underline{\varepsilon} R_{g2} \text{ (for all } p_{cm} > 0)$$

Some of these will be eliminated by future "umbrella groups" of future "sequential prime products". The complexities of this process are quite simple once the patterns are seen a number of times while working with this process. Elimination of some "prime candidates" is due to the fact that future "sequential prime products" will contain larger prime numbers that were not present in a current "sequential prime product" and its associated "repetition group". As a result, there will be "prime candidates" that will be correctly added to the existing "sequential prime product" that will give a result that is not a true prime (in the traditional factoring view of primes, these numbers would be numbers with primes as factors that are larger than the primes used to generate the current "sequential prime product").

The eventual elimination of these non-primes coincides with the fact that not all of the "prime candidates" in the present "umbrella group" will remain after being eliminated by future "umbrella groups", which have the benefit of larger prime numbers in their "repetition groups" and associated "sequential prime products". But it will become clear that future generations of "umbrella groups" would not be complete if it were not for the non-prime contributors in the ancestor "umbrella groups" which are later eliminated. This is a strange but essential fact in generating the primes by this method. Numerous examples of this principle will be given below. Another way of looking at this is brought from wave theory, where succeeding waves ride on top of and alter preceding waves. It is essential to understand that all of the "umbrella groups" begin at 0 and extend with periodicity S_{gn} to infinity, with each successive "wave" modifying the entire spectrum of prior "umbrella groups" with repeating patterns.

The "repetition group" $R_{g2} = \{2,3\}$ has "repetition group length" = 6. Adding and subtracting 1 to 6 gives 7 and 5, the first "prime candidates" in the "umbrella group" U_{g2} of the "sequential prime product" $S_{pp\,2} = 6$. Since 2 and 3 are members of the already stabilized primes used to generate 6, neither they nor any member of their "repetition group" R_{g2} can be added or subtracted to 6 to generate new "prime candidates". Starting with 5 (the first "prime candidate" of the previous "umbrella group" U_{g1} that is not a member of the "repetition group" R_{g2}) the complete set of "prime candidates" is calculated in the table below. Note that 25 and 35 are generated in this group and will be eliminated by the next "umbrella group", which will have 5 as a member of the next "repetition group". Note however that 25 and 35 are still used to generate other potential "prime candidates", but only in the current "umbrella group" generation process. Adding 6 to 25 and 35 gives rise to 31 and 41 which will turn out to be primes; after being eliminated in the next "umbrella group", 25 and 35 will no longer be used to generate potential primes. Now it is becoming clear why the development of the primes requires starting with 0 and 1 and proceeding in a set sequence. The periodicity of this "umbrella group" is 6 (e.g. every other number differs by 6), in both the positive and negative directions, staring with the value $S_{pp\,2} = 6$ (for all values > 0).

6 ± 1 = 7, 5
5 = 11, 1
7 = 13, n/a
11 = 17, n/a
13 = 19, n/a
17 = 23, n/a
19 = 25, n/a
23 = 29, n/a
25 = 31, n/a
29 = 35, n/a
31 = 37, n/a
35 = 41, n/a
37 = 43, n/a
41 = 47, n/a
43 = 49, n/a
...
...

Note again that this process continues to infinity and this constitutes the "umbrella group" U_{g2} of the "sequential prime product" $S_{pp\,2} = 6$. The patterns seen here will repeatedly emerge and fade in many modified forms as future generations of primes are created and future "umbrella groups" are generated that will modify this "umbrella group". Notice that already the prime numbers in the negative direction and towards 0 from prior "sequential prime products" in ancestor "umbrella groups" are stabilizing and will not change due to any future generation of primes. In this group, all numbers less than $9 = 3^2$ have been "stabilized" and will not be affected by any future "umbrella groups". The stabilized primes to this point are 2,3,5 and 7.

Note that since 5 is the largest prime number of the next "sequential prime product", there are sufficient "stabilized" prime numbers to assure that this algebraic process will proceed. The generalized topic of "stabilizing primes" is the topic of one of the yet unreleased accompanying papers to this first of four papers. The number 49 is generated above and it will continue to be a "prime candidate", being eliminated by the second future generation "umbrella group" which will have the benefit of 7 in its "sequential prime product".

Note that the already determined prime numbers 2 and 3 that were used to generate the "sequential prime product" $S_{g2} = 6$ do not appear in the table above. So the entire list of "stabilized" primes to this point consists of the union of the stabilized primes of all prior "umbrella groups". This concept will become more significant as the topics such as primes density, etc are investigated in the other papers of this series. The "umbrella group" U_{g2} is a subset of the prior "umbrella group" U_{g1}, with the terms 9, 15, 21, 27 ... being eliminated" as they are members of the "repetition group" R_{g2} as required by the "umbrella group" definition.

Next generate the subsequent "sequential prime product" S_{g3} and its related "umbrella group" U_{g3}.

Using 2, 3 and 5 generate the next "sequential prime product", and follow the generation rules for creating the "umbrella group" of "prime candidates".

$$U_{g3} = \Pi_1^{\,3}\, p_i \pm 1 = 30 \pm 1 = 31{,}29 \text{ and } \Pi_1^{\,3}\, p_i \pm p_{cm}\,,$$
$$p_{cm}\ \varepsilon\ U_{g2} \text{ and } p_{cm}\ \underline{\varepsilon}\ R_{g3} \ \text{(for all } p_{cm} > 0)$$

$30 \pm\ \ 1 = 31{,}29$
$7 = 37{,}23$
$11 = 41{,}19$
$13 = 43{,}17$
$17 = 47{,}13$
$19 = 49{,}11$
$23 = 53,\ 7$
$29 = 59,\ 1$
$31 = 61$, n/a
$37 = 67$, n/a
...
...

The already stabilized primes 2, 3 and 5 (which comprise the members of this "sequential prime product") are not present in the above list. The periodicity of this "umbrella group" is $S_{pp\,3} = 30$, repeating the same "prime candidate" pattern to infinity.

The decreasing primes discovered by subtracting the "prime candidates" gives the list to 0 and confirms the list already determined for "sequential prime product" $S_{pp\,2} = 6$. 25, 35 and all other members of the "repetition group" R_{g3} have been "eliminated" in this "umbrella group" that has the benefit of the prime number 5. It is important to understand that the potential "prime candidates" 25, 35, etc. are not being eliminated because they have factors other than 1 and themselves (as defined by the traditional factorization method used to determine if numbers are prime or not), they are eliminated because they are members of the "repetition group" R_{g3}. The "prime candidate" 49 will be eliminated with the next "umbrella group", which will contain 7 as one of the members of the "repetition group" and its associated "sequential prime product" $S_{g4} = 210$. But also note that 49 must be used in the above list as it will produce the prime number $30 + 49 = 79$, just as 25 and 35 were used above to generate primes 31 and 41 in the "umbrella group" of $S_{pp\,2} = 6$. If these "prime candidates" were not used to create the above "umbrella group", then the "umbrella group" would not be complete and would not be then modified correctly by the next "umbrella group". Once again, since 49 will be eliminated in the "umbrella group" of the next "sequential prime product" 210 (which will have the benefit of 7), 49 will cease to be used at that point and will no longer be used to add to "sequential prime products" to generate future "prime candidates". These same rules are applied as the entire "umbrella group" U_{g3} is generated to infinity.

At this point the principles of symmetry, reciprocity and completeness of the prime numbers are well illustrated. Symmetry results from the addition and subtraction of known "prime candidates" from the "sequential prime products". The definition of "umbrella group" by its very nature guarantees this symmetry. Reciprocity arises from the fact that the primes generated in the negative direction from a given "sequential prime product" coincide with the primes generated in the positive direction from an ancestor "sequential prime product". Reciprocity further is a property of all future (positive direction) primes in the umbrella groups of all present and future "umbrella groups". The completeness of the prime number system and closure as a complete autonomous number system (a closed group) should be apparent at this point as these properties continue to infinity. The formal proof of this is the topic of one of the other three papers associated with this paper, but sufficient information has been presented to show how this complex proof proceeds. The concept of ancestor and parent/child relationships should also be apparent as these "generation rules" continue to infinity. The strange and repetitive patterns of primes are also apparent as the future generations of primes are reflected in ancestor "umbrella groups" and modified by the periodic "waves" of future "umbrella groups". The "Wave Theory" of primes is also a topic of the associated papers. Examples will be given below.

Further Examples of Prime Generation and Prime Patterns

Sufficient information has now been presented to allow readers to continue and generate all future primes and "umbrella groups". A few further examples will be given to clarify some of the statements made throughout this paper.

In the "sequential prime product" based on the "repetition group" of R_{g4} = {2,3,5,7}, $S_{pp\,4}$ = 210 is used in the generation of its "umbrella group" of "prime candidates". 210 + 1 = 211, which is prime. However 210 – 1 = 209, which is not prime, and will be eliminated by the next "umbrella group" which will include 11 in its "repetition group". 11 is the first prime of $U_{g(4-1)} = U_{g3}$ and not a member of R_{g4}, so 210 - 11 = 199 which is a true prime, however 210 + 11 = 221 which is a non-prime that will not be "eliminated" with the next "umbrella group", but will in fact have to wait until the second future "umbrella group" which will include 13 in its "repetition group". As already noted, the process of elimination becomes more complex as future "umbrella groups" are generated, but the fundamental rules remain the same. The generalized proof for when members of an umbrella group become "stabilized", and are therefore guaranteed to be true primes, is the topic of one of the related papers. A simple subset of that proof is used for the benefit of simplicity in this paper. Remember that even the non-prime "prime candidates" in an "umbrella group" are essential to the complete generation of all primes. This leads to the topic of density of primes, prime patterns and other topics beyond the scope of this first of 4 papers.

The primes added to $S_{pp\,4}$ = 210 which will eventually prove to be true primes are: 1, 13, 17, 19, 23, 29, 31, 41, 47, 53, 59, 61, 67, 71, 73, 83, 97 and so on. Likewise, the primes subtracted from 210 that will eventually prove to be true primes are: 11, 13, 17, 19, 29, 31 and so on. It is a relatively simple exercise to complete this list by first establishing the entire "umbrella group" U_{g4} of the "sequential prime product" 210 and its subsequent modification by future "umbrella groups" until the above mentioned "stabilized" list is achieved.

Going to the next "sequential prime product" $S_{pp\,5}$ = 2310 the prime numbers which are added to generate the true primes of this "umbrella group" turn out to be 1, 23, 29, 31, 37, 41, 47, 61, 67, 71, 73, 79, 83 and so on. The prime numbers which are subtracted from 2310 to generate the true primes in the negative direction are 1, 13, 17, 23, 29, 37, 41, 43, 59, 67, 71, 73, 79, 87, and so on. . It is a relatively simple exercise to complete this list by first establishing the entire "umbrella group" U_{g5} of the "sequential prime product" 2310 and its subsequent modification by future "umbrella groups" until the above mentioned stabilized list is achieved.

Any prime number can be calculated from any "sequential prime product" $S_{pp\,n}$. For example, select 1427, which is a prime number. From 210 add 1217, which is a prime number (210 + 1217 = 1427). From 2310 subtract 883, which is a prime number (2310 – 883 = 1427). Or one can add prime numbers as in the example 2310 + 1051 = 3361; all of these being prime numbers.

One must be careful since there are some values of the differences between prime numbers and "sequential prime products", which will not result in true prime numbers. These are due to the fact that the non-prime differences are "prime candidates" of the "umbrella group" U_{gn} associated with the assocated "sequential prime product" $S_{pp\,n}$. As an example, 331 is a prime number. 331- 210 = 121, but 121 is not a prime number. This is explained by the present method of viewing primes by the fact that 121 is in the "umbrella group" $S_{pp\,4} = 210$. 121 will be "eliminated" from use in generating prime numbers in the "umbrella group" U_{g5}. So until that point any number that is a member of a higher order "repetition group" such as 11 (or other higher order primes for that matter) may be found in the prime tables. Likewise, the "sequential prime product" $S_{pp\,5} = 2310$ finds a similar example. The number 1409 is prime. 2310 – 1409 = 901 which is not prime (e.g. the difference between $S_{pp\,5}$ and the prime number 1409 is not prime). But 901 will not be eliminated until the "umbrella group" U_{g7} associated with the "sequential prime product" $S_{pp\,7} = 510{,}510$ and its related "repetition group" R_{g7} containing the prime number 17. The operator guarantees that every false prime will identified and only the true and complete list of primes will remain up to the criteria set previously (for all numbers less than the square of the largest member of the "repetition group").

There are strange patterns that emerge such as the sequence of primes starting with 907: 907, 911, 919, 929, 937, 941, 947, 953, 967 and 971. These seemingly follow the pattern of primes 7, 11, 13, 17, 19, 23, 29, etc (but with some numbers missing). This is what was noted earlier. These patterns seem to arise and then fade as new patterns emerge. Likewise the same pattern emerges again at 1,511 and continues in the 1,600's but again fades away: 1511, 1523, 1531, 1543, 1549, 1553, 1559, 1567, 1571, 1579, 1583, 1597, 1601, 1607, 1609, 1613 and 1619. At this point the "wave" begins to fade and new patterns emerge. Note that 900 and 1500 are both multiples of $S_{pp\,3} = 30$. Similar patterns will be obvious around all multiples of the infinite list of "sequential prime products", all of which give rise to waves with wavelength of the same values.

In spite of these unusual patterns, the entire spectrum of primes can be calculated given the methods described in this paper and understood in terms of the periodic "waves" of "prime candidates" and ultimately "stabilized" prime numbers generated in the succession of "umbrella groups" with repetition "wave lengths" equal to the "sequential prime products" = $S_{pp\,n}$. Additionally, one can be assured that all of the primes can be identified without exception, that is, the generated group of prime numbers is complete.

Associating prime numbers follows from this discussion. For example, adding and subtracting the prime number 7 to any multiple of 30 will generate certain primes. The following table shows related primes in this family of primes based on wavelength 30.

$30 \pm 7 =$ 23,30
$60 \pm 7 =$ 53,67
$90 \pm 7 =$ 83,97
$120 \pm 7 =$ 113,127

However 150 - 7 gives 143, which will be eliminated by the "repetition group" containing 11. 150 + 7 = 157 will be discovered to be a true prime and will not be eliminated by any future application of the operator. One can also view all of these numbers from the vantage point of $S_{pp\,4} = 210$ (or any future "sequential prime product" $S_{pp\,n}$). For example 210 – 187 = 23 where 187 is not a true prime but is in the "umbrella group" of 210 (e.g. $S_{pp\,4} = 210$ still has 11 in its "umbrella group"; 187 will not be eliminated until the operator works on the "umbrella group" with "repetition group" $R_{g5} = \{2,3,5,7,11\}$). 210 – 173 = 37 where both 173 and 37 are true primes. Even though 210 ± 7 will not produce a prime (since 7 is in the "repetition group" of 210), some future products of 30 (e.g. $n \times 30 \pm 7$) which are greater than 270 will produce primes such as 270 + 7 = 277. However there will never be another prime number of the form $n \times 210 \pm 7$ (the wave based on wavelength 210, n = 1,2,3....).

The point is that the placement of prime numbers in the prime number table is not random and the prime numbers can now be understood relative to any of the "sequential prime products" in terms of waves and relationships to all the other prime numbers originating with the numbers 0 and 1. The essential basis of this is the operator defined in the Alternative Definition of Prime Numbers. The concept of taking a primitive set of numbers such as 0 and 1 and generating the entire set of prime numbers is a tribute to this powerful mathematical tool and holds promise for future applications with more complex building blocks in conjunction with more complex generator function operators. This has applications in the realms of genetics, quantum mechanics, and complete analytical solutions to N-Body problems such as in celestial mechanics as well as muli-dimensional topology. The solution to the Riemann Hypothesis may be closer using this as a basis for understanding the true nature of the prime numbers. The most illusive and long-standing mathematical problem in both Mathematics and Physics, the method of directly calculating the prime numbers, is found in this solution.

Additional Comments

There are numerous other topics that will be mentioned briefly that have been developed but which are outside the scope to this first paper. These include the following.

It will become apparent that "natural" bases of number systems for working with prime numbers will be 2, 6, 10, 30, etc. These will allow various unique views of the prime numbers.

The new Alternative Definition of Prime Numbers has an associated proof for the infinite number of primes. Also, the nature of the "umbrella groups" gives rise to studies of prime numbers using wave analysis.

The density of primes and theoretical work on proofs such as the number of primes between the squares of numbers is viewed differently. It now may make more sense to talk about the number of primes between the squares of successive prime numbers rather than the squares of all numbers.

The computational efforts to arrive at prime numbers will be very different from the traditional method of factorization. The amount of computation required based on the factorization method has diminishing returns on every aspect of the process. As the potential prime numbers get larger, the process is complicated by the fact that almost every number has to be tested, the number of factors increases as the numbers get larger, the number of primes needed to be included in the test grows, and the density of primes goes down, so there are increasingly more negative result tests. It is a process of diminishing returns. The new method of locating primes shows that the prime lists are stabilized relatively quickly, the candidates are never "tested" but are directly calculated from an ever decreasing list of "prime candidates". Additionally, patterns and relationships can be programmed into the search for identification of designated prime associations.

The study of the symmetry of primes is a topic that will take on a new future as the relationships between past and future prime groups are studied.

Factorization of any number now can be defined as the difference between any number and any of the "sequential prime products" $S_{pp\,n}$. This is because the "sequential prime products" are based on the prime numbers. For example, the number 81 is subtracted and added to $S_{pp\,4} = 210$. $210 - 81 = 129$. Both 81 and 129 have 3 as factors. $210 + 81 = 291$. 291 and 81 both have common factors.

Using this method, any number can be tested to be prime or not but with the requirement that one needs to know the members of the "umbrella groups" associated with the "sequential prime product" being used in the test.

There is now a new deeper understanding of the prime numbers that offer solutions to other more complex problems. Many of these are the topics of the other as yet unreleased papers of this series.

Conclusion

A new set of concepts was presented in the first of four papers. This first paper defined the concepts of "repetition patterns", repetition groups", sequential prime products", "generation rules" and "elimination rules", the concepts of symmetry, reciprocity and completeness and demonstrated the prime number system to be a closed group based only on the operations of addition and subtraction, beginning with only the numbers 0 and 1. To affect this, a new Alternative Definition of Prime Numbers was given. These give rise to the concepts of "umbrella groups", "generations" such as "ancestor", "successor", "parent" and "child" groups of primes. "Prime candidates" are generated in infinite groups and not just one at a time. By application of this method, successive "umbrella groups" are generated which eliminate certain "prime candidates" in periodic "waves" of periodicity or "wave length" $S_{pp\,n}$, while leaving other "prime candidates" which eventually become "stabilized", and remain as the true prime numbers.

Sufficient information was given to show the direction the other papers must take to prove these properties for the general case of all primes, including the completeness of the group of primes as closed group, not requiring multiplication or any other numbers to generate the entire set of prime numbers.

The primes are an elegant self-generating closed group of numbers, and not simply the errant lost orphans of an orderly multiplication table of non-primes.

XIX. OTHER SOURCES OF INFORMATION

Below I have listed some of the more popular books available today in print classified as Easy – Medium - Hard. I recommend you remain focused on credentialed sources for information on the prime numbers as you search for more information. But as you read these, be ever aware that this material was generated before the understanding of the prime numbers was released in this book and related papers. The entire motivation for the complex mathematical analysis such as "The Riemann Hypothesis" was *because there was no solution to directly calculate or examine the prime numbers or their distribution amongst the other non-prime numbers.*

As already noted, the mathematical community gave up over 100 years ago any attempts at finding a rhyme or reason to the prime numbers. More recently some have even claimed a proof that the prime numbers are "random", that is, having no discernable mathematical structure. Clearly that is not the case. Not only are there amazing patterns, but also the prime numbers are a closed family of numbers unrelated to the other numbers. They *ARE NOT* simply the lost errant children of the orderly multiplication table of non-primes.

So you will find the sentiment repeated in everything you read, that the solution to these highly complex (and many yet unsolved) mathematical problems regarding prime numbers are of paramount importance. Now one can directly calculate the density of prime numbers because you can directly calculate the primes based on the information given in this book.

BUT, because the mathematics of the prime numbers (based on "The Riemann Hypothesis", etc.) has become such an integral part of higher mathematics, physics and other branches of study, and because so much is based on these fundamental equations, they today stand as EXTREMELY important in the world of physics and mathematics. They are no doubt closer to solutions as a result of the present work, but their solutions may now be related to the true understanding of the prime number system rather than being a method of generalizing about prime number properties as they have been in the past. This is an

In searching for more information on the prime numbers and related problems, many people will search on the Internet. The good web sites are generally associated with mathematical associations and university mathematics departments. As the popularity of the prime number topic grows with the general public, there may very well be an increase in the number of the usual "Internet Imposters" trying to use the topic of prime numbers to gain attention, with everything from predicting the future to who knows what. I have even seen where Internet imposters have tried to create names that might imply a mathematical background or claim to have degrees, which they do not have, and upon examination, they are simply quacks. As with anything on the Internet, it is a case of "buyers beware". Since the Internet is a changing environment, and since this book will be published and printed into the future, I am not recommending any Internet sites, simply follow the simple rules in your search. The official web site for this book is **www.calculateprimes.com**

Easy books:

Stalking the Riemann Hypothesis: The Quest to Find the Hidden Law of Prime Numbers by Dan Rockmore (2006)

The Riemann Hypothesis: The Greatest Unsolved Problem in Mathematics by Karl Sabbagh (2003)

The Millennium Problems: The Seven Greatest Unsolved Mathematical Puzzles of Our Time by Keith J. Devlin - Good chapter on the Riemann Hypothesis.

The Music of the Primes: Searching to Solve the Greatest Mystery in Mathematics by Marcus du Sautoy (2003)

Medium books:

Prime Obsession: Bernhard Riemann and the Greatest Unsolved Problem in Mathematics by John Derbyshire (2004) - Lots of equations accompanied by very understandable verbal explanations. Once this book is digested then Edwards book can be tackled.

Hardest book:

Riemann's Zeta Function by Harold M. Edwards (2001) - contains an English translation and explanation of Riemann's original paper.

XX. 0 & 1 – NEGATIVE AND COMPLEX PRIMES

There are a number of topics that are not required for this text but which may be of interest to mathematicians and physicists attempting to see beyond the rudimentary process of calculating the prime numbers. There are fundamental questions that arise since the new process of directly calculating prime numbers starts with 0 and 1. An argument can be made for inclusion of 0, 1, negative numbers and complex numbers in an expanded list of prime numbers.

Using both the traditional and new definitions of prime numbers, and based on the possible applications in solutions to numerical problems, the following discussion opens the door for inclusion of these numbers in an expanded list of prime numbers. Once the insight into the usefulness of negative prime numbers is realized, the next logical step is to include 0 and 1 as the "connecting" numbers bridging the positive primes with the negative primes. The new definition of prime numbers developed in this book produces the exact same list of prime numbers as the traditional factorization method, but has the additional implication that there is a rational for including 0, 1, the negative primes $-p_k$ and primes in complex forms $p_n + p_m i$, where i is the square root of -1, and the "sequential prime products" $S_{pp\,n}$ have negative and complex forms, being used in the same manner.

Are 0, 1, negative values of the prime numbers $-p_k$ and complex prime numbers valid primes using the traditional factorization method of defining prime numbers? I am not attempting to answer these questions in this book and DVD, I am only raising the possibility as I am already using them in my own research. There are of course many opinions on the use of 1 as a prime number already in existence.

$\lim_{\Delta x \to 0} \Delta x / \Delta x = 1$ and $0 / 1 = 0$, therefore, 0 is prime.

$1 / 1 = 1$ and 1 has only itself as a factor, therefore 1 is prime.

$-p_i / -p_i = 1$ and $-p_i$ has only itself as a factor (for all $-p_i < 0$)

$p_n + p_m i / p_n + p_m i = 1$, and $p_n + p_m i$ has only itself as a factor.